Rural–Urban Water Struggles

Rural–Urban Water Struggles compiles diverse analyses of rural–urban water connections, discourses, identities and struggles evolving in the context of urbanization around the world.

Departing from an understanding of urbanization as a process of constant making and remaking of multi-scalar territorial interactions that extend beyond traditional city boundaries and that deeply reconfigure rural–urban hydrosocial territories and interlinkages, the chapters demonstrate the need to reconsider and trouble the rural–urban dichotomy. The contributors scrutinize how existing approaches for securing urban water supply – ranging from water transfers to payments for ecosystem services – all rely on a myriad of techniques: they are produced by, and embedded in, specific institutional and legal arrangements, actor alliances, discourses, interests and technologies entwining local, regional and global scales. The different chapters show the need to better understand on-the-ground realities, taking account of inequalities in water access and control, as well as representation and cultural-political recognition among rural and urban subjects.

Rural–Urban Water Struggles will be of great use to scholars of water governance and justice, environmental justice and political ecology.

This book was originally published as a special issue of *Water International*.

Lena Hommes is an Associated Researcher with the Water Resources Management Group of the Department of Environmental Sciences at Wageningen University, the Netherlands. Her research focuses on hydrosocial territories, rural–urban water relations, politics of water infrastructure projects, water justice struggles and socio-environmental transformations in Latin America.

Rutgerd Boelens is a Professor of Water Governance and Social Justice in the Department of Environmental Sciences at Wageningen University, the Netherlands, and a Professor of Political Ecology of Water in Latin America with CEDLA (Centre for Latin American Research and Documentation) at the University of Amsterdam, the Netherlands. He is also a Visiting Professor at the Catholic University of Peru (Department of Social Sciences), Lima, Peru, and the Central University of Ecuador (Faculty of Agricultural Sciences), Quito, Ecuador, and coordinates the international Justicia Hídrica / Water Justice Alliance. He is also the main scientific editor of the book series *Water and Society*, bringing together various Latin American publishing houses. His work includes articles and books focusing on water rights, water grabbing, hydrosocial territories, legal pluralism, cultural politics, governmentalities and social mobilizations in Latin America and Spain.

Leila M. Harris is a Professor with the Institute for Resources, Environment and Sustainability and with the Institute for Gender, Race, Sexuality and Social Justice at the University of British Columbia (UBC), Vancouver, Canada. At UBC, she is also a member of the EDGES research collaborative and serves as the co-Director for the Program on Water Governance (PoWG). Her research focuses on the social, political and equity dimensions of environment and resource issues, with particular interest in water politics and governance. Recent projects have included focus on water access and governance in Ghana, South Africa and Canada. Internationally, she is also involved with the International WaTERS Research and Training Network as well as the Household Water Insecurity Experiences (HWISE) Research Network.

Gert Jan Veldwisch is an Associate Professor with the Water Resources Management Group of the Department of Environmental Sciences at Wageningen University, the Netherlands. His research focuses on the practices, policies and politics of agricultural water management, including farmer-led irrigation development, water grabbing, waste water use in agriculture, agrarian change and issues around water justice. Currently, he works mostly in Southern and Eastern Africa, but he has also worked in Latin America and Central Asia.

Routledge Special Issues on Water Policy and Governance

Edited by: Cecilia Tortajada (IJWRD) – Institute of Water Policy, Lee Kuan Yew School of Public Policy, NUS, Singapore
James Nickum (WI) – International Water Resources Association, France

Most of the world's water problems, and their solutions, are directly related to policies and governance, both specific to water and in general. Two of the world's leading journals in this area, the *International Journal of Water Resources Development* and *Water International* (the official journal of the International Water Resources Association), contribute to this special issues series, aimed at disseminating new knowledge on the policy and governance of water resources to a very broad and diverse readership all over the world. The series should be of direct interest to all policy makers, professionals and lay readers concerned with obtaining the latest perspectives on addressing the world's many water issues.

Integrated Water Management in Canada
The Experience of Watershed Agencies
Edited by Dan Shrubsole, Dan Walters, Barbara Veale and Bruce Mitchell

Groundwater and Climate Change
Multi-Level Law and Policy Perspectives
Edited by Philippe Cullet and Raya Marina Stephan

OECD Principles on Water Governance
From Policy Standards to Practice
Edited by Aziza Akhmouch, Delphine Clavreul, Sarah Hendry, Sharon Megdal, James Nickum, Francisco Nunes-Correia and Andrew Ross

Urban Resilience to Droughts and Floods
The Role of Policies and Governance
Edited by Cecilia Tortajada, James Horne and Larry Harrington

Politics and Policies for Water Resources Management in India
Edited by M. Dinesh Kumar

Rural–Urban Water Struggles
Urbanizing Hydrosocial Territories and Evolving Connections, Discourses and Identities
Edited by Lena Hommes, Rutgerd Boelens, Leila M. Harris and Gert Jan Veldwisch

For more information about this series, please visit: https://www.routledge.com/series/WATER

Rural–Urban Water Struggles

Urbanizing Hydrosocial Territories and Evolving Connections, Discourses and Identities

Edited by
Lena Hommes, Rutgerd Boelens, Leila M. Harris and Gert Jan Veldwisch

Routledge
Taylor & Francis Group
LONDON AND NEW YORK

IWRA

First published 2020
by Routledge
2 Park Square, Milton Park, Abingdon, Oxon, OX14 4RN

and by Routledge
52 Vanderbilt Avenue, New York, NY 10017

Routledge is an imprint of the Taylor & Francis Group, an informa business

British Library Cataloguing-in-Publication Data
A catalogue record for this book is available from the British Library

ISBN13: 978-0-367-37004-6

Typeset in Minion Pro
by codeMantra

Publisher's Note
The publisher accepts responsibility for any inconsistencies that may have arisen during the conversion of this book from journal articles to book chapters, namely the inclusion of journal terminology.

Contents

Citation Information

The chapters in this book were originally published in *Water International*, volume 44, issue 2 (February 2019). When citing this material, please use the original page numbering for each article, as follows:

Introduction

Rural–urban water struggles: urbanizing hydrosocial territories and evolving connections, discourses and identities
Lena Hommes, Rutgerd Boelens, Leila M. Harris and Gert Jan Veldwisch
Water International, volume 44, issue 2 (February 2019) pp. 81–94

Chapter 1

Water crisis through the analytic of urban transformation: an analysis of Bangalore's hydrosocial regimes
Michael Goldman and Devika Narayan
Water International, volume 44, issue 2 (February 2019) pp. 95–114

Chapter 2

The rural–urban equity nexus of Metro Manila's water system
Philamer C. Torio, Leila M. Harris and Leonora C. Angeles
Water International, volume 44, issue 2 (February 2019) pp. 115–128

Chapter 3

Hydrosocial territories in the context of diverse and changing ruralities: the case of Cochabamba's drinking water provision over time
Paul Hoogendam
Water International, volume 44, issue 2 (February 2019) pp. 129–147

Chapter 4

Colonizing rural waters: the politics of hydro-territorial transformation in the Guadalhorce Valley, Málaga, Spain
Bibiana Duarte-Abadía and Rutgerd Boelens
Water International, volume 44, issue 2 (February 2019) pp. 148–168

Chapter 5

The political construction and fixing of water overabundance: rural–urban flood-risk politics in coastal Ecuador
Juan Pablo Hidalgo-Bastidas and Rutgerd Boelens
Water International, volume 44, issue 2 (February 2019) pp. 169–187

Chapter 6

Upsetting the apple cart? Export fruit production, water pollution and social unrest in the Elgin Valley, South Africa
Matthijs Wessels, Gert Jan Veldwisch, Katarzyna Kujawa and Brian Delcarme
Water International, volume 44, issue 2 (February 2019) pp. 188–205

Chapter 7

Hydrosocial territories, agro-export and water scarcity: capitalist territorial transformations and water governance in Peru's coastal valleys
Gerardo Damonte and Rutgerd Boelens
Water International, volume 44, issue 2 (February 2019) pp. 206–223

Chapter 8

Payment for ecosystem services in Lima's watersheds: power and imaginaries in an urban-rural hydrosocial territory
Sonja Bleeker and Jeroen Vos
Water International, volume 44, issue 2 (February 2019) pp. 224–242

Chapter 9

Conclusions: Evolving connections, discourses and identities in rural–urban water struggles
Lena Hommes, Gert Jan Veldwisch, Leila M. Harris and Rutgerd Boelens
Water International, volume 44, issue 2 (February 2019) pp. 243–253

Notes on Contributors

Leonora C. Angeles is an Associate Professor at the Institute for Gender, Race, Sexuality and Social Justice, and the School of Community and Regional Planning at the University of British Columbia, Vancouver, Canada. Her continuing research and interests are on community and international development studies and social policy; participatory planning and governance; participatory action research; and the politics of transnational feminist networks, women's movements and agrarian issues, particularly in the Southeast Asian region.

Sonja Bleeker is a Research Consultant in the Water Resources Management Group in the Department of Environmental Sciences at Wageningen University, the Netherlands. She is also an independent consultant in watershed management, and a Lecturer in Disaster Risk Management at Van Hall Larenstein University of Applied Sciences, Leeuwarden, the Netherlands. She has a strong interest in the urban water cycle and inequities in access and control over water in urbanized watersheds.

Rutgerd Boelens is a Professor of Water Governance and Social Justice in the Department of Environmental Sciences at Wageningen University, the Netherlands, and a Professor of Political Ecology of Water in Latin America with CEDLA (Centre for Latin American Research and Documentation) at the University of Amsterdam, the Netherlands. He is also a Visiting Professor at the Catholic University of Peru (Department of Social Sciences), Lima, Peru, and the Central University of Ecuador (Faculty of Agricultural Sciences), Quito, Ecuador, and coordinates the international Justicia Hídrica / Water Justice Alliance. He is also the main scientific editor of the book series *Water and Society*, bringing together various Latin American publishing houses. His work includes articles and books focusing on water rights, water grabbing, hydrosocial territories, legal pluralism, cultural politics, governmentalities and social mobilizations in Latin America and Spain.

Gerardo Damonte is a Senior Researcher in the Department of Social Sciences at the Pontifical Catholic University of Peru (PUCP), Lima, Peru. He also carries out research in the field of natural resources at the Group for the Analysis of Development (GRADE), Peru. Here, he leads projects in three interrelated fields: extractive industries and rural society, local knowledge and climate change, and territories and social movements.

Brian Delcarme is a Lecturer at the Centre for Water and Sanitation Research at Cape Peninsula University of Technology, Cape Town, South Africa.

Bibiana Duarte-Abadía is a Ph.D. Researcher at the Centre for Latin American Research and Documentation (CEDLA) at the University of Amsterdam, the Netherlands. She has conducted research regarding hydrosocial territories transformations focused on the political ecology of water in Colombia, Mexico, and Spain, specifically in fluvial systems and highland ecosystems.

Michael Goldman is an Associate Professor in the Department of Sociology and the Institute for Global Studies at the University of Minnesota, Minneapolis, USA. His research areas of expertise are transnational sociology of development, environment, financial institutions, knowledge/power, expertise, land politics and global urbanisms/cities.

Leila M. Harris is a Professor with the Institute for Resources, Environment and Sustainability and with the Institute for Gender, Race, Sexuality and Social Justice at the University of British Columbia (UBC), Vancouver, Canada. At UBC, she also is a member of the EDGES research collaborative and serves as the co-Director for the Program on Water Governance (PoWG). Her research focuses on the social, political and equity dimensions of environment and resource issues, with particular interest in water politics and governance. Internationally, she is also involved with the International WaTERS Research and Training Network as well as the Household Water Insecurity Experiences (HWISE) Research Network.

Juan Pablo Hidalgo-Bastidas works at the Centre for Latin American Research and Documentation (CEDLA), and at the Department of Geography, Planning and International Development Studies at the University of Amsterdam, the Netherlands, and at the Faculty of Agricultural Sciences at the Central University of Ecuador, Quito, Ecudaor.

Lena Hommes is an Associated Researcher with the Water Resources Management Group of the Department of Environmental Sciences at Wageningen University, the Netherlands. Her research focuses on hydrosocial territories, rural–urban water relations, politics of water infrastructure projects, water justice struggles and socio-environmental transformations in Latin America.

Paul Hoogendam is a Water and Climate Change Program Coordinator at Consulting Engineers Salzgitter GmbH in Germany. He has been affiliated with the Department of Environmental Sciences at Wageningen University, the Netherlands.

Katarzyna Kujawa is a Lecturer and Researcher in the Environmental Technology Group at Wageningen University and Research, the Netherlands. She also researches at LeAF, an independent research and consultancy organization. Her expertise is in environmental engineering, environmental technology, aerobic treatment, anaerobic digestion, waste treatment, waste water treatment and water supply.

Devika Narayan is a PhD Student in the Department of Sociology at the University of Minnesota, Minneapolis, USA. Her doctoral project focuses on the structural changes underway in India's software industry – the strategic, organizational and workforce-related issues that emerge in the face of disruptive technological shifts. More generally, her academic interests lie in the areas of global political economy, industrial/economic sociology and labor studies.

Philamer C. Torio is a Senior Lecturer in Integrated Water Systems and Governance at the IHE Delft Institute for Water Education, the Netherlands, where he coordinates and teaches courses on water partnerships and financing. He has senior management experience in project and corporate finance, project development, business management and public-private partnerships.

Gert Jan Veldwisch is an Associate Professor with the Water Resources Management Group of the Department of Environmental Sciences at Wageningen University, the Netherlands. His research focuses on the practices, policies and politics of agricultural water management, including farmer-led irrigation development, water grabbing, waste water use in agriculture, agrarian change and issues around water justice. Currently, he works mostly in Southern and Eastern Africa, but he has also worked in Latin America and Central Asia.

Jeroen Vos is an Assistant Professor in the Water Resources Management Group in the Department of Environmental Sciences at Wageningen University, the Netherlands. He teaches courses on water governance, institutional reform in the water sector, and irrigation and development. His academic interest is in the combination of thorough technical analysis with institutional, legal, social and economic concepts.

Matthijs Wessels is a Lecturer and a PhD Candidate in the Water Resources Management Group in the Department of Environmental Sciences at Wageningen University, the Netherlands. His fields of expertise are irrigation management, wastewater irrigation, water reuse, water quality and urban-rural linkages.

INTRODUCTION

Rural–urban water struggles: urbanizing hydrosocial territories and evolving connections, discourses and identities

Lena Hommes, Rutgerd Boelens, Leila M. Harris and Gert Jan Veldwisch

Introduction

In 2007, for the first time in human history, the world's population became more urban than rural – a trend that is expected to increase in the coming decades, reaching 66% by 2030. While being 'powerhouses of economic growth and catalysts for inclusion and innovation', cities face severe challenges to provide adequate services, including drinking water and sanitation (UN-Habitat, 2016, p. iii). Correspondingly, one of the Sustainable Development Goals is focused on cities and achieving access to affordable housing and basic services for all by 2030 (United Nations, 2017, p. SDG11). This directly links with other goals, such as SDG 6, which focuses on universal access to safe drinking water and adequate sanitation.

Despite all the international attention on urban development, and the changing character of urban life and infrastructure, critical analyses of the entanglement of cities and their surrounds are often missing. Further, the rural–urban binary that scholarly and policy discussions often highlight can be misleading, as the concurrent rise of peri-urban development and the periodic and cyclical migration of people between urban, extra-urban and rural areas challenge such a neat dichotomy. The central focus of this special issue is on water-related analysis that explicitly links urban locales to the broader surrounds. The authors herein examine a range of urban–rural and peri-urban linkages in terms of the flows, pathways and governance of water, and the ways in which urban processes trigger political, technological, economic and cultural shifts in rural / peri-urban / urban linkages.

It is clear that the development of cities and urban conglomerations closely relates to the dynamics and relations in rural areas, and centrally depends on the appropriation and transformation of diverse rural socio-environments. In the past, such linkages have been largely theorized through concepts such as 'water, ecological, or urban footprints', with geographic concepts such as the urban hinterland, or with perspectives that examine 'the rural' as extraction zones for city wealth and the dumping grounds for urban waste disposal. Here, we build on and enrich the range of discussions and analytics to further theorize urban–rural water connections. To do so, we particularly call attention to the interrelatedness of material, political, social and discursive aspects of rural–urban water (dis)connections. As we outline below, hydraulic infrastructure projects, policies, laws and water-related imaginaries and discourses all connect rural and urban spaces – and thus attendant hydrosocial designs and configurations. This also includes attention to intersecting flows of water, knowledge, money, ideas and people, as processes that all require analytics that link rural and urban hydrogeographies (or, engaging other terminology, waterscapes [Swyngedouw, 1997], hydrosocial cycles [Linton & Budds, 2014], or hydrosocial territories [Boelens, Hoogesteger, Swyngedouw, Vos, & Wester, 2016]). While some of the earlier concepts and approaches invite attention to rural–urban connections and linkages through quantitative analysis of resource requirements (as with the footprints approach), this special issue combines a range of analyses that further political ecology and hydrosocial approaches to illustrate and consider such connections. As such, our contributions help clarify that urban-rural changes in the water realm relate not only to important biophysical, climatological and material dynamics and linkages (as with the ecological footprint concept) but also rely on ideological, economic and political connections and linkages. We discuss how urban and rural livelihoods and lifeways are constructed relationally. Given our reliance on political ecology and associated socio-natural, techno-political and cultural-discursive frameworks, our contribution similarly advances approaches related to disputes over socio-ecological interventions, distribution and governance, as well as equity and justice considerations important for urban-rural water transformations. We also relate these concepts to issues of (disputed) water culture(s) and identities, examining the ways in which cultural politics often importantly shape urban–rural relations and mutual (mis)recognition. This combination of themes has only begun to be explored to date (it is most prevalent in earlier work on 'extraction' or 'sacrifice' zones, as noted above).

Water provides a particularly useful and interesting lens through which to understand and think through urban/rural and urban/extra-urban dynamics and pathways. As a flow resource, water moves from rural spaces, through cities, and out again. On the way, water literally takes on characteristics and 'imprints' from the diverse locales it passes through. It also links people and places in ways that make the connections between sites clear – often in ways that are hidden in analyses that take the urban–rural boundary to be fixed or distinct. Thus, upstream pollution (e.g., from a mining site) affects the suitability of the water source for use in an urban area, or can affect the specific treatment processes that are required (or urban and peri-urban pollution can pose particular risks for agricultural water uses; see Wessels, Veldwisch, Kujawa, & Delcarme, 2019, this issue). Similarly, urban residents benefit from forested and protected catchments upstream, and there has been increasing recognition of these connections with programmes such as payment for ecosystem services (PES) to protect water sources that serve urban populations – i.e., payments to upstream farmers to avoid practices that pollute shared waters or to reward them for pursuing practices that conserve and protect water sources. In a similar way, after water leaves urban areas, there is a clear and indelible imprint of the features of cities, from industrial pollution to storm runoff that might be associated with urban soil sealing. In all of these ways, water links and connects urban, rural and peri-urban areas, transcends such divides and also defies the characterization of these very categories. In this volume, we explore such dynamics to explicitly consider the interfaces and complex movements, flows and dynamics that connect and disrupt locales and the boundaries between them. We are also interested in exploring the 'boundary work' or boundary-making processes that performatively serve to cast and entrench rural / urban / peri-urban distinctions as fixed, or meaningful (Jones, 2009), despite the range of considerations (including water flows) that undermine and disrupt such categorization.

There are other important examples of how water illustrates, embodies and underscores these linkages and connections. For example, water from rural areas is extracted and piped to supply urban houses and industrial sites; irrigated food crops are imported to cities from rural areas (as 'virtual water'); and watersheds are reshaped through the construction of hydropower plants for energy production to serve urban and industrial growth (Heynen, Kaika, & Swyngedouw, 2006; Kaika, 2006). Even though the exact pathways through which nature is transformed to become part of urban accumulation and consumption differ across contexts, all cities are embedded in broader socio-natural and political systems, flows and territories that extend far beyond their geographical boundaries (Swyngedouw, 1997). It is clear that urbanization is not a territorially discrete and self-contained phenomenon but involves the constant making and remaking of multiscalar networks and interactions (Harris, 2015). In this way, city and countryside are at once connected, but at the same time the 'rural' also remains importantly differentiated from the 'urban' (Angelo & Wachsmuth, 2014). For example, municipal drinking water utilities might extend territorial claims over highland water resources, thus (technologically and political-administratively) interlinking urban water demands with rural territories and livelihoods; and, at the same time, producing particular and very differentiated water access and control patterns between 'the rural' and 'the urban' (Hommes & Boelens, 2017). To provide a very different example of the ways that the urban is consolidated in relation to the rural, urban dwellers often insist on particular imaginaries of what it means to be an 'urban citizen', at times invoking distinct relationships with water as key to distinguishing what it means (or should mean) to live in urban spaces. For instance, residents in Buenos Aires, Argentina, insist that appropriate sanitation is important for them, as they now live in urban areas and left behind the low-tech toilets they associate with rural life (Morales, Harris, & Öberg, 2014), while residents in townships around Cape Town similarly distinguish between life in the townships and their earlier rural lifeways in relation to various modes of water access (with linked implications for willingness to pay for water; see Rodina & Harris, 2016). In a similar manner, as various articles in this volume show, in many cases, urban-biased identity politics and modernity discourses legitimize transferring water from the countryside to provide water services to advanced city life (see e.g. Duarte-Abadía & Boelens 2019; Hoogendam, 2019; Wessels et al., 2019).

The contributions to this special issue aim to critically analyze these water-related connections between city and countryside, including peri-urban areas as the interstitial zones between them. Doing so allows us to explore the ways urbanization reconfigures adjacent rural territories and power relationships (and vice versa), the broader hydro-social dynamics at play, or the changing scales and pathways of water in these interlinked systems. We also ask whether, how and with what consequences rural water resources become affected and water-based rural livelihoods and identities transformed because of measures taken to meet growing urban water (or food and energy) demands. Likewise, we explore how power dynamics have differentiated effects both within and between rural and urban settings, which go beyond uniform concentrations of water in the direction of cities. Finally, we explore the potential for inclusive and democratic urban–rural relations and water governance, and related equity dimensions of shifting dynamics, based on case studies from around the world. For example, Bleeker and Vos (2019, this issue) show the necessity to take into consideration the complex social, economic and historical contexts and water realities, to not reinforce rural–urban inequalities through new urban water security projects in Peru. Wessels et al. (2019, this issue) call attention to drawbacks in decentralization and democratization efforts in South Africa.

Inspiration for this issue came from discussions between researchers of the International WaTERS Network (wwww.international-waters.org) and the Water Justice Alliance (*Justicia Hídrica*, www.justiciahidrica.org), held in 2016 in Wageningen, the Netherlands, in conjunction with the international PE-3C Conference (Political Ecologies of Conflict, Capitalism and Contestation). The outcome was an understanding that it was time to bring rapidly changing rural–urban relations and the linked, newly emerging socio-environmental justice issues to the agenda. In this introduction to the special issue we discuss the importance of a political-ecology approach to exploring the rural–urban interface and evolving hydro-social relations and propose some conceptual building blocks. In a separate, concluding contribution, we draw together the specific insights offered by the articles in this special issue.

Towards a political ecology of rural–urban hydrosocial relations

Using a political-ecology perspective for engaging with hydro-social transformations in urbanization processes and rural-urban metabolisms, this special issue builds on, and extends, scholarly work on urban water development and resource governance more broadly. Various authors have pointed to the importance of studying rural–urban linkages and how we can understand urban development and expansion, fundamentally, as a reconfiguration of adjacent rural territories. Essential groundwork was laid by William Cronon (1991), who showed in *Nature's Metropolis* how the urban growth of Chicago transformed surrounding areas, imposing on the land 'a new geography of second nature in which the market relations of capital reproduced themselves in an elaborate urban-rural hierarchy that would henceforth frame all human life in the region' (p. 378). Urbanization is a driving force behind environmental change, reconfiguring rural and rural-urban socio-economic, political and ecological realities in important ways (Heynen et al., 2006). Likewise, scholars such as Blackbourn (2006), Kaika (2006), Mumford (1961, 1971) and Swyngedouw (1997) suggest that pursuits of urban modernity are based on the conquest and disciplining of nature, with connections across diverse geographies.

These insights compel us to study hydro-social dynamics in the context of the interlinkages and interdependence between rural and urban spaces, repoliticizing transformations of rural–urban interfaces and scrutinizing what they mean, and for whom. As political ecology entails an analytical focus on power relations and inequalities, an urban-rural political-ecological focus invites attention precisely to the processes that produce inequalities in resource access, representation, cultural-political recognition, and hierarchization of knowledges among rural and urban subjects (see the analysis of the rural–urban equity nexus in Metropolitan Manila by Torio, Harris, & Angeles, 2019; as well as Goldman &

Narayan, 2019; Hidalgo-Bastidas & Boelens, 2019; and Hoogendam, 2019, all in this volume). Such an approach examines how inequities emerge in relation to territorial (re) configurations and relationships (Forsyth, 2003; Fraser, 2000; Heynen et al., 2006; Neumann, 2005). In that way, supply-side responses to urban water requirements, such as mega-hydraulic water transfers, PES schemes and 'water for all' discourses, need careful and critical analysis to uncover the inequalities and environmental injustices that are often inscribed in such projects (see e.g. Goldman, 2007, 2011; Joy, Kulkarni, Roth, & Zwarteveen, 2014; Radonic, 2017; Veldwisch, Franco, & Mehta, 2018). This could provide information to policy makers to help make urban water supply approaches more sustainable and equitable, and for civil society organizations and grass-roots federations to advocate needed shifts in understandings and policy approaches.

A helpful conceptual notion for understanding hydro-geographical and techno-political transformations is that of 'hydrosocial territories' (Boelens et al., 2016), understood as contested imaginary and dynamically materializing networks that entwine biophysical, technological, social and political structures, practices, actors, meanings, discourses and knowledges in which water is a central constituent. In this notion, territories are understood as disputed socio-natural arrangements, consisting of interlaced material, social and symbolic components. As such, hydrosocial territories are not a value-free, neutral construct shaped through an open-ended democratic process but imbued with power politics: 'politically speaking, territory is the socio-materially constituted and geographically delineated organization and expression of and for the exercise of political power' (Swyngedouw & Boelens, 2018, p. 117; see also Harris, 2012). This approach does not see territories as spaces delimited by fixed physical boundaries but focuses on how people continuously engage in boundary-making practices. These may be rooted, for example, in culturally embedded, place-based social practices or identities (Escobar, 2001), or equally in globally promoted paradigms. An example is the construction and framing of the river basin as the appropriate water governance boundary (Budds & Hinojosa Valencia, 2012; Cohen & Harris, 2013; Warner, Wester, & Bolding, 2008). In hydrosocial territories, constructing water infrastructure is an example of both boundary making and connection making, channelling water flows to link (or disrupt) places, people and practices (Hommes, Boelens, Duarte-Abadía, Hidalgo, & Hoogesteger, 2018).

In the process of hydro-territorial (re)configuration, a multiplicity of actors compete to realize territorial orderings according to their interests and worldviews. Given the large diversity of actors within one and the same space, this is a highly contested and ongoing process: Territories are never entirely and forever consolidated, but are constantly renegotiated. One important cause of the struggles are contested imaginaries of what territory is and what it should be, as well as the contestation over the actual materialization of these territorial components and compositions. Thus, on the one hand, the socio-territorial imaginaries have symbolic and social components that envision certain types of development and modernity, sustained by moral judgments and normative statements, for instance about good governance, responsible and irresponsible citizens, and efficient and inefficient water uses (Goldman, 2007; Harris, 2012; Hommes, Boelens, & Maat, 2016). On the other hand, imaginaries turn into important geographical forces that change the environment once materialized (Swyngedouw, 2015). Because of the divergent views and interests, in everyday reality multiple territorial projects coexist and interact. 'Territorial pluralism', therefore, is the dynamic outcome of diverse overlapping, interacting and conflicting ideas and territories in one and the same geographical-political space (Hoogesteger, Boelens, & Baud, 2016).

This way, massive hydraulic engineering projects transferring water to cities are often surrounded and justified by discourses of progress and development, contrasting modern, 'enlightened' cities and urban citizens with corollaries of rural backwardness (Harvey & Knox, 2012; Hommes & Boelens, 2018). In effect, local ways of living, rural knowledge and identities, as well as the possible contestations by affected rural populations of water transfer plans, are often delegitimized (Hidalgo-Bastidas, Boelens, & Isch, 2018). For instance, in his impressive historical study about the making of modern Germany, David Blackbourn (2006) shows how conquering and taming nature through reclaiming marshes, straightening rivers and constructing large dams went hand in hand with the modernist designs of the twentieth century, and was linked to racist ideologies and imperial projects in ways that fostered urbanization and the ordering of rural 'ungoverned territories'.[1]

To realize territorial imaginaries and projects, actors or actor alliances mobilize a variety of resources, ranging from financial and political resources to discursive strategies and truth claims. For example, as Hidalgo-Bastidas and Boelens observe in their article (2019, this issue), in the case of the Ecuadorian Chone mega-dam project, truth claims about flood risk are an integral part of the realization of the territorial project, which is imagined by an urban-political elite to safeguard urban modernity. Thus, territorial struggles are not only about natural resources but also – and not less importantly – about 'meanings, norms, knowledge, decision-making authority, and discourses' (Boelens et al., 2016, p. 8; cf. Baviskar, 2007; Swyngedouw, 2009). While other concepts are also useful to highlight complex social, political, ecological and hydrologic processes and their dynamic interplay (e.g., waterscapes and the hydrosocial cycle – see Linton & Budds, 2014 – or multiscalar analysis of water and conflict geographies, see Harris, 2002), the key is to highlight shifting configurations, economies, agro-ecologies, power dynamics and outcomes linked to ongoing changes and networked configurations across urban and rural geographies (e.g., Damonte & Boelens, 2019; Goldman & Narayan, 2019; Hoogendam, 2019, all in this volume).

Evolving rural–urban connections, technologies and social mobilization

So what kind of socio-territorial interlinkages evolve from urbanization processes taking place in globalizing, political and economically stratified societies that are competing over water resources that are becoming increasingly scarce? In the following, we identify several dynamics that play an essential role in transforming rural–urban hydrosocial relations. These dynamics are not isolated from each other, but interconnected and at times overlapping. It is also not our aim to enumerate *all* rural–urban water-related dynamics, but to highlight key themes of interest for the case studies in this issue.

First, there are water transfers that reallocate water from rural to urban areas via massive hydraulic systems, composed of dams, tunnels and canals that are literally cast in concrete or made of other sturdy, immobilizing materials. Gupta and van der Zaag (2008, p. 30) determine that in the year 2000, inter-basin water transfers contributed 14% of all water withdrawals worldwide. Even though these numbers might not be entirely up to date, rural–urban infrastructure-based water transfers have remained a popular option for politicians faced with water supply challenges for urban populations. This is not necessarily because they are the most cost-efficient or the only possible solution, but also because they are connected to powerful underlying imaginaries, utopian visions and expected political gains (Hommes & Boelens, 2017). Growing urban populations, ongoing hydrologic changes associated with climate change, and other shifts suggest that these large-scale water transfers are likely to remain important into the future (witness the recent issues in Cape Town).

The development of water transfers and other infrastructure is not only a material intervention but is simultaneously a deeply social and often highly political matter. Both the technical design and the construction process are informed by designers' and sponsors' interests, identities, particular knowledges and worldviews, meaning that infrastructural designs tend to reflect dominant governance and political positions (Mollinga & Veldwisch, 2016; Oberborbeck Andersen, 2016; Winner, 1980). Far from being an outcome of objective technical evaluation, technological projects reveal forms of rationality and governmentality (Hommes et al., 2016; Larkin, 2013). Accordingly, their effects are social and symbolic as well as material. As Jensen and Morita (2017, p. 6) find, infrastructure holds 'the potential capacity to do such diverse things as making new forms of sociality, remaking landscapes, defining novel forms of politics, reorienting agency, and reconfiguring subjects and objects, possibly all at once'. In other words, 'infrastructure reconfigures relations between people and states, spirits, forms of knowledge and nature' (p. 4; see also Morales et al., 2014).

Numerous studies in political geography and the philosophy of technology have made inquiries into these dimensions. They have analyzed, for example, how technology mediates politics by shaping interactions, power relations, self-consciousness and social awareness (Kaika, 2006; Meehan, 2013; Mollinga & Veldwisch, 2016); how water infrastructure serves as conduit of power that cultivates or delimits state power (Boelens, 2015; Harris, 2012; Meehan, 2014); how citizenship is mediated and claimed through water supply systems (Illich, 1986; Kooy & Bakker, 2008; Rodina & Harris, 2016); how water rights, property relations and governance arrangements are embedded in hydraulic infrastructure (Boelens & Vos, 2014); and how inequality is reproduced through structures, rules and practices related to that infrastructure (Anand, 2015; Redfield & Robins, 2016; Sanchis-Ibor, Boelens, & García-Mollá, 2017).

Similarly, the social contents and positioning of infrastructure and technologies in hydro-social territory-making should be understood as being appropriated, changed and negotiated in their day-to-day use and according to their physical and social surroundings. For example, water flows may not behave as envisaged by engineers, and affected water users may change how technologies are used in unforeseen ways (Sanchis-Ibor et al., 2017; Veldwisch, Bolding, & Wester, 2009). These insights, in consequence, require rural-urban political ecology to study infrastructure-based water transfers in detail, asking what changes and for whom as water moves from rural to urban areas. The dynamics at play in forging and enabling these transfers are also subject to analysis and consideration, from their planning and surrounding imaginaries, through the construction process, to their effects and appropriation in everyday lives over time.

Besides the 'traditional', infrastructure-based water transfers, a new approach for ensuring water supply to cities is PES. Global players such as the US-based Nature Conservancy and the World Wildlife Fund are promoting PES schemes in different forms as potential win-win approaches in urbanizing regions in Latin America, as well as in Europe and elsewhere (Bennett, Carroll, & Hamilton, 2013). These schemes and their advocates propose that upstream communities be rewarded in some way (usually financial or in-kind) for conserving areas in the upper watershed and changing their land- and water-based activities to presumably more sustainable models, so that water availability for downstream urban areas is secured (Goldman, Benitez, Calvache, & Ramos, 2010; Gutman, 2006). Critics caution that these new mechanisms are based on market principles, regarding them as yet another form of accumulation by dispossession ('accumulation by conservation' – Büscher & Fletcher, 2015, p. 273 – see also Cavanagh, 2018) or as overriding local land and water institutions and management relationships (Boelens, Hoogesteger, & Rodriguez de Francisco, 2014).

In any case, what is notable is that new mechanisms for ensuring water supply, such as PES schemes, do not simply rely on laws, regulations and governmental force but are based on the assumption that each individual is acting out of a self-interested rationality and that the right incentives are enough to motivate appropriate behaviour (Fletcher, 2010). In that regard, PES approaches can be understood as neoliberal governmentality endeavours that establish incentive structures motivating water users to behave according to neoliberal worldviews and market-environmentalist ideas about watershed conservation and natural resources management (Duarte-Abadía & Boelens, 2016). Institutions and discourses help establish such incentive structures, which, at the same time, also sanction non-participation and brand non-participating water users as abnormal and backward – often coinciding with the label 'rural' (Rodríguez de Francisco & Boelens, 2016; see also Bleeker & Vos, 2019, this volume).

Connected to that is the question whether there are newly evolving discourses surrounding the expansion of cities' claims on resources. Has the content of discourses supporting water supply projects moved from references to modernity and the need to overcome nature to 'develop', to references to the human right to water, Sustainable Development Goals, and environmental concerns? With increasing attention and focus on the human right to water, the earlier millennium development goals, and now the SDGs, we have seen a concurrent shift from water available for agriculture and rural areas to prioritization of urban needs (Harris, Rodina, & Morinville, 2015; Mirosa, & Harris, 2011; Torio et al., 2019, this volume). This is linked to discursive shifts that support drinking water and domestic needs, as well as ecological needs, over food crops or other productive uses, with clear implications for food security, livelihoods and other dimensions of urban–rural water linkages (Birkenholtz, 2016; Goff & Crow, 2014; Molle & Berkoff, 2006; Shah & Zerriffi, 2017). Even with a growing number of examples along these lines, from India, Peru, the Philippines, and elsewhere, there is still much to be known about how discourses around urban water supply or flood protection projects reconfigure rural areas and create potentially unequal water management arrangements. This is a problem especially for those negatively affected by urban projects, who are often also people with lesser economic, political and social power, as their concerns and queries remain unheard. This said, there are, of course, important equity considerations within urban and rural areas as well. As Torio et al. (2019, this volume) make clear, the considerable inequities in water access in the Metro Manila area appear to be worsened by ongoing inefficiencies in the urban water system. In a similar manner, Wessels et al.

(2019, this volume) point to the fact that labourers working in the agro-export industry are living in precarious housing and sanitation conditions in the city of Grabouw, South Africa, as the municipality cannot keep up with urban expansion. These disadvantaged populations are also at times blamed for the deteriorating water quality affecting agricultural production in the nearby Elgin Valley. Equity considerations become linked to rural areas in other ways as well. As Hoogendam (2019, this volume) shows in his analysis of how diverse and changing ruralities in the Cochabamba area have been differently affected by the city's search for new water resources, while valley peasants have succeeded in defending their water access and limiting municipal groundwater exploitation, highland peasants have not managed to avert negative impacts of urban water supply infrastructure in their territories.

The variety of rural–urban territorial transformations is enormous. They may be triggered by growing urban claims on local water resources, by global agro-product market demands, by calls for ecological conservation or by specific ideologies, as the contributions to this volume make clear. In all these transformations, the interests and socio-territorial projects and imaginaries of different water user groups compete within the same space and time. Each actor is striving to realize a certain kind of territorial project that contains hydraulic-physical as well as socio-economic and cultural aspects – and that is not easily reconcilable with other actors' projects. In consequence, the transformation of rural-urban hydrosocial territories is a process of continuous (re)negotiations and struggles that go far beyond technical questions about where water should flow to or not. These struggles are also about equity in participation and decision making, representation and recognition of identities, values, beliefs and knowledges (Fraser, 2000; Schlosberg, 2004; cf. Jackson, 2018). At the same time, the place-bound particularity of these struggles, and correspondingly the importance of territory and space in negotiating socio-economic positions, is central. In the words of Harvey (1990, p. 419): 'To challenge what the place might be, physically as well as socially, is to challenge something fundamental in the social order.' At the same time, the territorial imaginaries and projects of different actor groups might clash, or contain fierce environmental injustices, but nevertheless may not lead to struggles or contestations. Cleaver (2018), for example, points to the need for more attention to everyday relations of water injustice that are characterized by acceptance, compromise and adjustment.

Concluding remarks

In this introduction we have pointed out the relevance of studying changing relations of water use and access in the context of rural–urban (dis)connections. We have laid out a conceptual foundation for studying cases and have pointed out core themes of interest, drawing from both long-standing and more recent literatures. We have emphasized the importance of studying water relations and dynamics in the interconnections between resources, rules and regulations, discourses, identities and authority. We are particularly interested in how materiality and power dynamics interact in the constitution and change of social-material networks around water. The contributions to this issue are diverse in their geographical location (including case studies from the Philippines, South Africa, India, Spain and Latin America), as well as in their specific themes within the field of rural–urban water struggles. In our concluding discussion (pages 243–253), we draw together the insights that we have gained from these contributions. We have learned a lot in working with these authors and bringing together this special issue, and hope that a careful reading offers all of our audiences reason for pause and reflection, and opportunities for critical insight.

Note

1. Phalangist/Francoist urbanization projects in Spain (e.g., Boelens & Post Uiterweer, 2013; Swyngedouw, 2015; Duarte-Abadía & Boelens, 2019, in this volume), apartheid-driven territorial planning in Southern Africa (Rodina & Harris, 2016. Mahlanya, Ziervogel, & Scott, 2016; cf .Wessels et al., 2019, in this volume), and racial-neocolonial hydropolitics prevailing in many Latin American cases (e.g., Boelens, 2015) show similar ideologies and practices. Nazi Germany took it to historical extremes. Blackbourn (2006) shows how modern-versus-backward identity dichotomies were actively 'inscribed in', and presumably could be 'read off from', the German land/waterscape. 'Ordered' and 'rational' white-Aryan urbanistic land and waterscapes were distinguished from wastelands populated by 'parasitic' Jews and 'lazy' Slavs, and once the Nazi regime had 'cleansed' the landscape (massively deporting 'inferior races'), ethnic Germans were resettled on their territories.

Disclosure statement

No potential conflict of interest was reported by the authors.

Funding

This work forms part of the activities of the Water Justice/Justicia Hídrica Alliance (www.justiciahi drica.org) and was supported by a grant of the Social Sciences and Humanities Research Council for the International WaTERS research and training network [Grant no: 890-2013-0020].

References

Anand, N. (2015). Leaky states: Water audits, ignorance and the politics of infrastructure. *Public Culture, 27*(2), 305–330. doi:10.1215/08992363-2841880

Angelo, H., & Wachsmuth, D. (2014). Urbanizing urban political ecology: A critique of methodological cityism. *International Journal of Urban and Regional Ressearch*, 16–27. doi:10.1111/1468-2427.12105

Baviskar, A. (2007). *Waterscapes. The cultural politics of a natural resource.* Delhi: Permanent Black.

Bennett, G., Carroll, N., & Hamilton, K. (2013). Charting new waters: State of watershed payments 2012. *Washington, DC.* Retrieved from http://www.ecosystemmarketplace.com/reports/sowp2012

Birkenholtz, T. (2016). Dispossessing irrigators: Water grabbing, supply-side growth and farmer resistance in India. *Geoforum, 69*(C), 94–105. doi:10.1016/j.geoforum.2015.12.014

Blackbourn, D. (2006). *The conquest of nature: Water, landscape and the making of modern Germany.* London: Norton.

Bleeker, S., & Vos, J. (2019). Payment for ecosystem services in Lima´s watersheds: Power and imaginaries in an urban-rural hydrosocial territory. *Water International, 44*(2), 224–242. doi:10.1080/02508060.2019.1558809

Boelens, R. (2015). *Water, power and identity. The cultural politics of water in the Andes.* London: Routledge.

Boelens, R., Hoogesteger, J., & Rodriguez de Francisco, J. C. (2014). Commoditizing water territories: The clash between andean water rights cultures and payment for environmental services policies. *Capitalism Nature Socialism*, 1–19. doi:10.1080/10455752.2013.876867

Boelens, R., Hoogesteger, J., Swyngedouw, E., Vos, J., & Wester, P. (2016). Hydrosocial territories: a political ecology perspective, *Water International*, 41(1), 1–14, doi:10.1080/02508060.2016.1134898

Boelens, R., & Post Uiterweer, N. C. (2013). Hydraulic heroes: The ironies of utopian hydraulism and its politics of autonomy in the Guadalhorce Valley, Spain. *Journal of Historical Geography*, *41*, 44–58. doi:10.1016/j.jhg.2012.12.005

Boelens, R., & Vos, J. (2014). Legal pluralism, hydraulic property creation and sustainability: The materialized nature of water rights in user-managed systems. *Current Opinion in Environmental Sustainability*, *11*, 55–62. doi:10.1016/j.cosust.2014.10.001

Budds, J., & Hinojosa-Valencia, L. (2012). Restructuring and rescaling water governance in mining contexts: The co-production of waterscapes in Peru. *Water Alternatives*, *5*(1), 119–137.

Büscher, B., & Fletcher, R. (2015). Accumulation by conservation. *New Political Economy*, *20*(2), 273–298. doi:10.1080/13563467.2014.923824

Cavanagh, C. J. (2018). Critical ecosystem infrastructure: Governing the forests– Water Nexus in the Kenyan Highlands. In R. Boelens, T. Perreault, & J. Vos (Eds.), *Water Justice* (pp. 302–315). Cambridge, UK: Cambridge University Press.

Cleaver, F. (2018). Everyday water injustice and the politics of accommodation. In R. Boelens, T. Perreault, & J. Vos (Eds.), *Water justice* (pp. 246–258). Cambridge, UK: Cambridge University Press.

Cohen, A., & Harris, L. (2013). Performing watersheds: Performativity and the production of scale. In R. Rose-Redwood & M. Glass (Eds.), *Performativity, politics and the production of social space*. (pp. 226-250) London: Routledge.

Cronon, W. (1991). *Nature's metropolis*. New York: W.W. Norton & Company.

Damonte, G., & Boelens, R. (2019). Hydrosocial territories, agro-export and water scarcity: Capitalist territorial transformations and water governance in Peru's coastal valleys. *Water International*, *44*(2), 206–223. doi:10.1080/02508060.2018.1556869

Duarte-Abadía, B., & Boelens, R. (2016). Disputes over territorial boundaries and diverging valuation languages: The Santurban hydro-social highlands territory in Colombia. *Water International*, *41*(1), 15–36. doi:10.1080/02508060.2016.1117271

Duarte-Abadía, B., & Boelens, R. (2019). Colonizing rural waters. The politics of hydro-territorial transformation in the Guadalhorce Valley, Málaga, Spain. *Water International*, *44*(2), 148–168. doi:10.1080/02508060.2019.1578080

Escobar, A. (2001). Culture sits in places: Reflections on globalism and subaltern strategies of localization. *Political Geography*, *20*, 139–174. doi:10.1016/S0962-6298(00)00064-0

Fletcher, R. (2010). Neoliberal environmentality: Towards a poststructuralist political ecology of the conservation debate. *Conservation and Society*, *8*(3), 171–181. doi:10.4103/0972-4923.73806

Forsyth, T. (2003). *Critical political ecology. The politics of environmental sciences*. London: Routledge.

Fraser, N. (2000). Rethinking recognition. *New Left Review*, *3*, 107–120.

Goff, M., & Crow, B. (2014). What is water equity? The unfortunate consequences of a global focus on 'drinking water'. *Water International*, *39*(2), 159–171. doi:10.1080/02508060.2014.886355

Goldman, M. (2007). How "water for all!" policy became hegemonic: The power of the world bank and its transnational policy networks. *Geoforum*, *38*(5), 786–800. doi:10.1016/j.geoforum.2005.10.008

Goldman, M. (2011). The birth of a discipline: Producing authoritative green knowledge, world bank-style. *Ethnography*, *2*(2), 191–217. doi:10.1177/14661380122230894

Goldman, M., & Narayan, D. (2019). Water crisis through the analytic of urban transformation: an analysis of Bangalore's hydrosocial regimes. *Water International*, *44*(2), 95–114. doi:10.1080/02508060.2019.1578078

Goldman, R. L., Benitez, S., Calvache, A., & Ramos, A. (2010). *Water funds: Protecting watersheds for nature and people*. Virginia: Arlington.

Gupta, J., & van der Zaag, P. (2008). Interbasin water transfers and integrated water resources management: Where engineering, science and politics interlock. *Physics and Chemistry of the Earth, Parts A/B/C*, *33*(1–2), 28–40. doi:10.1016/j.pce.2007.04.003

Gutman, P. (2006). PES: A WWF perspective. *Paper presented at the Conservation Finance Global Biodiversity Forum*, Critiba, Brazil.

Harris, L. M. (2002). Water and conflict geographies of the Southeastern Anatolia Project. *Society and Natural Resources*, *15*(8), 743–759. doi:10.1080/08941920290069326

Harris, L. (2012). State as socionatural effect: Variable and emergent geographies of the state in Southeastern Turkey. *Comparative Studies of South Asia, Africa and the Middle East*, *32*(1), 25–39. doi:10.1215/1089201X-1545345

Harris, L. (2015). Scalar politics, networks and power in water governance. In E. Norman, C. Cook, & A. Cohen (Eds.), *Negotiating water governance: Why the politics of scale matter* (pp. 226–250). London: Ashgate.

Harris, L., Rodina, L., & Morinville, C. (2015). Revisiting the human right to water from and environmental justice lens. *Politics, Groups and Identities*, *3*(4), 660–665. doi:10.1080/21565503.2015.1080619

Harvey, D. (1990). Between space and time: Reflections on the geographical imagination. *Annals of the Association of American Geographers*, *80*(3), 418–434.

Harvey, P., & Knox, H. (2012). The enchantments of infrastructure. *Mobilities*, *4*(4), 521–536. doi:10.1080/17450101.2012.718935

Heynen, N., Kaika, M., & Swyngedouw, E. (2006). Urban politial ecology: Politicizing the production of urban natures. In N. Heynen, M. Kaika, & E. Swyngedouw (Eds.), *In the nature of cities* (pp. 1–19). London: Routledge.

Hidalgo-Bastidas, J. P., & Boelens, R. (2019). The political construction and fixing of water over-abundance. Rural-urban flood risk politics in Coastal Ecuador. *Water International*, *44*(2), 169–187. doi:10.1080/02508060.2019.1573560

Hidalgo-Bastidas, J. P., Boelens, R., & Isch, E. (2018). Hydroterritorial configuration and confrontation: The Daule-Peripa multipurpose hydraulic scheme in Coastal Ecuador. *Latin American Research Review*, *53*(3), 517–534. doi:10.25222/larr.362

Hommes, L., & Boelens, R. (2017). Urbanizing rural waters: Rural-urban water transfers and the reconfiguration of hydrosocial territories in Lima. *Political Geography*, *57*, 71–80. doi:10.1016/j.polgeo.2016.12.002

Hommes, L., & Boelens, R. (2018). From natural flow to 'working river': Hydropower development, modernity and socio-territorial transformations in Lima's Rímac watershed. *Journal of Historical Geography*, *62*, 85–95. doi:10.1016/j.jhg.2018.04.001

Hommes, L., Boelens, R., Duarte-Abadía, B., Hidalgo, J. P., & Hoogesteger, J. (2018). Reconfiguration of hydrosocial territories and struggles for water justice. In R. Boelens, T. Perreault, & J. Vos (Eds.), *Water justice* (pp. 151–168). Cambridge, UK: Cambridge University Press.

Hommes, L., Boelens, R., & Maat, H. (2016). Contested hydrosocial territories and disputed water governance: Struggles and competing claims over the Ilisu Dam development in south-eastern Turkey. *Geoforum*, *71*, 9–20. doi:10.1016/j.geoforum.2016.02.015

Hoogendam, P. (2019). Hydrosocial territories in the context of diverse and changing ruralities: The case of Cochabamba's drinking water provision over time. *Water International*, *44*(2), 129–147. doi:10.1080/02508060.2019.1551711

Hoogesteger, J., Boelens, R., & Baud, M. (2016). Territorial pluralism: Water users' multi-scalar struggles against state ordering in Ecuador's highlands. *Water International*, *41*(1), 91–106. doi:10.1080/02508060.2016.1130910

Illich, I. (1986). *H20 and the waters of forgetfulness*. London: Marion Boyars.

Jackson, S. (2018). Water and indigenous rights: Mechanisms and pathways of recognition, representation, and redistribution. *WIREs Water*, *5*, e1314. doi:10.1002/wat2.1314

Jensen, C. B., & Morita, A. (2017). Introduction: Infrastructures as ontological experiments. *Ethnos*, *82*(4), 615–626. doi:10.1080/00141844.2015.1107607

Jones, R. (2009). Categories, borders and boundaries. *Progress in Human Geography*, *33*(2), 174–189.

Joy, K. J., Kulkarni, S., Roth, D., & Zwarteveen, M. (2014). Re-politicising water governance: Exploring water re-allocations in terms of justice. *Local Environment*, *19*(9), 954–973.

Kaika, M. (2006). Dams as symbols of modernization: The urbanization of nature between geographical imagination and materiality. *Annals of the Association of American Geographers, 96*(2), 276–301.

Kooy, M., & Bakker, K. (2008). Technologies of government: Constituting subjectivities, spaces, and infrastructures in colonial and contemporary Jakarta. *International Journal of Urban and Regional Ressearch, 32*(2), 375–391.

Larkin, B. (2013). The politics and poetics of infrastructure. *Annual Review of Anthropology, 42*, 327–343.

Linton, J., & Budds, J. (2014). The hydro-social cycle: Defining and mobilizing a relational-dialectical approach to water. *Geoforum, 57*, 170–180.

Mahlanya, L., Ziervogel, G. & Scott, D. (2016). Water, rights and poverty: An environmental justice approach to analysing water management devices in Cape Town. *Urban Forum, 27* (4), 363–382.

Meehan, K. (2013). Disciplining de facto development: Water theft and hydrosocial order in Tijuana. *Environment and Planning D: Society and Space, 31*, 319–336.

Meehan, K. (2014). Tool-power: Water infrastructure as wellsprings of state power. *Geoforum, 57*, 215–224.

Mirosa, O., & Harris, L. (2011). Human right to water: Contemporary challenges and contours of a global debate. *Antipode, 44*(3), 932–949.

Molle, F., & Berkoff, J. (2006). *Cities versus agriculture: revisiting intersectoral water transfers, potential gains and conflicts* (pp. 1–80). Colombo, Sri Lanka: Comprehensive Assessment Secretariat.

Mollinga, P., & Veldwisch, G. J. (2016). Ruling by canal: Governance and system-level design characteristics of large-scale irrigation infrastructure in India and Uzbekistan. *Water Alternatives, 9*(2), 222–249.

Morales, M., Harris, L., & Öberg, G. (2014). Citizenshit: The right to flush and the urban sanitation imaginary. *Environment and Planning A, 46*, 2816–2833.

Mumford, L. (1961). *The city in history: Its origins, its transformations, and its prospects*. New York: Harcourt, Brace and World.

Mumford, L. (1971). *The myth of the machine II. The pentagon of power*. New York: Harcourt, Brace and World.

Neumann, R. (2005). *Making political ecology*. New York: Routledge.

Oberborbeck Andersen, A. (2016). Infrastructures of progress and dispossession: Collective responses to shrinking water access among farmers in Arequipa, Peru. *Focaal, 74*, 28–41. doi:10.3167/fcl.2016.740103

Radonic, L. (2017). Through the aqueduct and the courts: An analysis of the human right to water and indigenous water rights in Northwestern Mexico. *Geoforum, 84*, 151–159. doi:10.1016/j.geoforum.2017.06.014

Redfield, P., & Robins, S. (2016). An index of waste: Humanitarian design, "dignified living" and the politics of infrastructure in Cape Town. *Anthropology Southern Africa, 39*(2), 145–162. doi:10.1080/23323256.2016.1172942

Rodina, L., & Harris, L. M. (2016). Water services, lived citizenship, and notions of the state in marginalised urban spaces: The case of Khayelitsha, South Africa. *Water Alternatives, 9*(2), 336–355.

Rodríguez de Francisco, J. C., & Boelens, R. (2016). PES hydrosocial territories: De-territorialization and re-patterning of water control arenas in the Andean highlands. *Water International, 41*(1), 140–156. doi:10.1080/02508060.2016.1129686

Sanchis-Ibor, C., Boelens, R., & García-Mollá, M. (2017). Collective irrigation reloaded. Re-collection and re-moralization of water management after privatization in Spain. *Geoforum, 87*, 38–47. doi:10.1016/j.geoforum.2017.10.002

Schlosberg, D. (2004). Reconceiving environmental justice: Global movements and political theories. *Environmental Politics, 13*(3), 517–540. doi:10.1080/0964401042000229025

Shah, S., & Zerriffi, H. (2017). Urban water demand, climatic variation, and irrigation-water insecurity: Interactive stressors and lessons for water governance from the Angat River basin (Philippines). *Water International, 42*(5), 543–567. doi:10.1080/02508060.2017.1342073

Swyngedouw, E. (1997). Power, nature, and the city. The conquest of water and the political ecology of urbanization in Guayaquil, Ecuador: 1880–1990. *Environment and Planning A*, *29*(2), 311–332. doi:10.1068/a290311

Swyngedouw, E. (2015). *Liquid power: Contested hydro-modernities in 20th century Spain*. Cambridge, MA: MIT Press.

Swyngedouw, E. (2009). The political economy and political ecology of the hydrosocial cycle. *Journal of Contemporary Water Research & Education*, 142, 56–60. doi:10.1111/jcwr.2009.142.issue-1

Swyngedouw, E., & Boelens, R. (2018). And not a single injustice remains": Hydro- territorial colonization and techno- political transformations in Spain. In R. Boelens, T. Perreault, & J. Vos (Eds.), *Water justice* (pp. 115–133). Cambridge, UK: Cambridge University Press.

Torio, P., Harris, L., & Angeles, L. (2019). The rural-urban equity nexus of Metro Manila's water system. *Water International*, *44*(2), 115–128. doi:10.1080/02508060.2019.1560559

UN-Habitat. (2016). *Urbanization and development: Emerging futures - World cities report 2016*. Nairobi, Kenya: UN-Habitat.

United Nations. (2017). Sustainable development goal 11. Retrieved from https://sustainabledevelopment.un.org/sdg11

Veldwisch, G. J., Bolding, A., & Wester, P. (2009). Sand in the engine: The travails of an irrigated rice scheme in Bwanje Valley, Malawi. *Journal of Development Studies*, *45*(2), 197–226. doi:10.1080/00220380802265587

Veldwisch, G. J., Franco, J., & Mehta, L. (2018). Water grabbing: Practices of contestation and appropriation of water resources in the context of expanding global capital. In R. Boelens, T. Perreault, & J. Vos (Eds.), *Water justice* (pp. 59–70). Cambridge, UK: Cambridge University Press.

Warner, J., Wester, P., & Bolding, A. (2008). Going with the flow: River basins as the natural units for water management? *Water Policy*, *10*(S2), 121–138. doi:10.2166/wp.2008.210

Wessels, M., Veldwisch, G. J., Kujawa, K., & Delcarme, B. (2019). Upsetting the Apple Cart? Export fruit production, water pollution, and social unrest in the Elgin Valley, South Africa. *Water International*, *44*(2), 188–205. doi:10.1080/02508060.2019.1586092

Winner, L. (1980). Do artifacts have politics? *Daedalus*, *109*(1), 121–136.

Water crisis through the analytic of urban transformation: an analysis of Bangalore's hydrosocial regimes

Michael Goldman and Devika Narayan

ABSTRACT

This paper explores intensified water crisis in Bangalore (or Bengaluru) in India by using the analytic of three hydrosocial regimes: the catchment-based regime, the hydraulic regime and the speculative urban regime. It uses a wide range of qualitative interviews, scientific reports and secondary sources to analyze shifting urban trajectories, agrarian relations and their interlinkages with water. Historical ruptures (in the realm of governance, urban growth and changing urban–rural dynamics) allow one to highlight the complex role of speculative logics that shape urban expansion and water scarcity.

Introduction

Water shortage has plagued Bangalore (or Bengaluru) in India for many decades, but it is only recently that government officials, journalists, non-governmental organizations (NGOs) and citizens have adopted the language of crisis. In 2017, news headlines reflected this anxiety:

> Dry fact: Bengaluru is paying the price for killing its water bodies. (*News Laundry*, 2017)
>
> Bengaluru water crisis: Karnataka faces severe scarcity, 160 of 176 taluks (or sub-districts) declared drought-hit. (*Financial Express*, 2017)
>
> City of burning lakes: experts fear Bangalore will be uninhabitable by 2025. (*The Guardian*, 2017)
>
> Water Crisis: Is Bengaluru headed for Day Zero? (*Times of India*, 2018)

In early 2018, reportage on the water problem again surged with a BBC (2018) report naming Bangalore as the second most likely city in the world (behind Cape Town) to run out of drinking water in the near future. The year 2018 saw the city's largest lake catch fire multiple times due to the abundant toxins and debris that were dumped in it. The previous year, the lake had been in the news for spewing large quantities of toxic foam onto the streets (Upadhya, 2017). Meanwhile, hydrologists, historians and geographers are also beginning to portray the water situation in India's cities as a crisis (Lele et al., 2013; Mehta, Goswami, Kemp-Benedict, Muddu, & Malghan, 2013;

Ranganathan, 2010). The shift in the discourse characterizing Bangalore's water situation, from water shortage to crisis, offers an opportunity for academic reflection (Prakashi, 2017; Raj, 2013a, 2013b; Ramachandra & Aithal, 2016a).

In this paper, we argue that the main features of Bangalore's urban transformation – growing from a mid-sized town into the 'Silicon Valley of Asia' with promises of global-city infrastructure and expansion – are rooted in distinct practices of urbanization that are coupled with *shifts in hydrosocial territorial use*. This notion of hydrosocial territories comes from Boelens, Hoogesteger, Swyngedouw, Vos, and Wester (2016) who use it to capture the *relational dimensions* central to the making and unmaking of water scarcities and crisis. Their conceptual themes of the centrality of 'territorialization' and the 'politics of scalar territorial reconfigurations' not only underline the significance of interactions among social uses of land and water but also lead us to consider relational and scalar dimensions. We build on their work by conceptualizing Bangalore's interactions with its hydrological resources in terms of *hydrosocial regimes*, or relatively stable periods where sources and distributions of water and their spatial and social features develop and yet become limited due to continual urban expansion fuelled by different engines of growth. Here, we take as our starting point political ecology's foundational argument that natural and social processes are co-constitutive, always influenced by the contexts of political economies and ecologies in which they are situated.

We examine Bangalore's water infrastructure through the lens of its three hydrosocial regimes: the catchment-based regime, the hydraulic regime and the speculative urban regime.

Like recent scholarship of the political ecology of water (Swyngedouw, 2009; Bakker, 2010; Budds, Linton, & McDonnell, 2014), we call for a 'repoliticization' of water such that we account for the relative and uneven distribution of water scarcity *and* abundance. We add to their approaches by making visible new sociopolitical pressures on an already fragile, crisis-oriented hydrosocial terrain, showing how these pressures result from growth-based and then speculative dynamics. For example, we demonstrate the growing importance of finance capital in its extractive capacity, first through debt financing of the state's water infrastructure, and then its funding of economically and ecologically unviable, speculative projects through its onerous-debt financing of developers and city agencies. Here, we discuss the prevalent *speculative urban* tendency to bank on rural land values continuing to increase and people's investments to rise, which is linked to the imaginary that the intense water woes in both the city and the countryside will resolve themselves without disruptions to the model of speculative urban growth and governance.

By delineating these trends as 'regimes' rather than straightforward chronologies, we want to underscore the point that it is not possible to separate colonial, state-capital and neoliberal eras of water use neatly since these hydrosocial dynamics are nested and overlapping. Instead, we prefer to stress key ruptures, events and conjunctures in which the political–ecological landscape changes dramatically (Sheppard et al., 2015). Our approach demonstrates how Bangalore's history of water struggles and crisis, when embedded in an understanding of shifting urban–rural relations and scalar power dynamics across (local–national–global) governance and economic structures, can reveal the central importance of hydrosocial territory-making and the ways its integrity can be undermined (Hommes, Boelens, Harris, & Veldwisch, 2019).

The analysis that follows is based on a combination of primary and secondary data sources. For the first two regimes we identify, the data come primarily from historical documents and reports, complemented by interviews by the authors of those works and other experts who are knowledgeable about Bangalore's water history. Our characterization of the third regime is based on our extensive research on the remaking of Bangalore as a 'global city'. Part of this research has involved conducting interviews (during the period 2007–16) with individuals working within the municipal water utility, Bangalore Water Supply and Sewerage Board (BWSSB), and with its hired consultants (e.g., Larsen and Tubro, PricewaterhouseCoopers), and current and retired administrators. We have also interviewed farmers, land brokers, real estate developers, water experts, water brokers, senior government officials, ecology researchers and low-level bureaucrats about themes related to water in Bangalore as part of a larger research project on the making of Bangalore into a global city, and use those interviews to inform this analysis. Although our larger project on speculative urbanism studies the transformation of the city as a whole, much of the data we present here focus on the area in the northern part of Bangalore, around the new international airport and information technology (IT) and biotech industrial zones. The last two data sources include published and unpublished studies and reports on the current water crisis, and a series of public events on water in Bangalore in which we participated, including two we co-organized in 2014 and 2016. Interviews with people who conducted action research and surveys on water use and agriculture in the early 1990s have given unique access to a historical timeline that has been corroborated by scholars of the region's ecological history.

The catchment-based regime (I)

Making the city possible

The first hydrosocial regime is characterized by an intricate network of people-made wells, lakes and tanks linked by channels and contoured embankments that were constructed by local residents, rulers and eventually the colonial government (Mathur & Da Cunha, 2006). We briefly discuss its evolution and explore the destruction and decline of this elaborate, decentralized system.

The 'catchment-based hydrosocial regime' has its origins in the era of the rule of the Maharaja of Mysore, expanded in the British era of urban development and continued with post-Independence public works projects. In their 2018 report, Mundoli, Manjunatha and Nagendra explain the history of the city of Bangalore as rooted in the construction of water infrastructure, dating back to 1537 AD, when Kempe Gowda I initiated the construction of Sampangi, Karanji and Agrahara lakes surrounding what is still the centre of the old city (Mundoli, Manjunatha, & Nagendra, 2018). As it is a plateau region with no perennial river and in the rain shadow of the Deccan Plateau, 900 metres above sea level, people constructed Bangalore's water infrastructure by using the undulating landscape, creating a wetlands environment of numerous water-holding facilities – called *kere*, lakes, irrigation tanks, reservoirs or ponds – some as large as 1.7 square miles.

Ramachandra, the most prominent scholar on the topic, calls the built-up ecosystem a 'wetlands treasure' of maintained forests, to create what was once admired as India's 'garden city' (Ramachandra et al., 2016c). When the British East India Company and later the British Crown took over the city, starting in 1799, the military settled in its cantonment and constructed a series of channel-linked, gravity-driven water tanks (e.g., Miller, Shoolay, Sankey and Ulsoor tanks) to supply ample water, at least for the cantonment and its surroundings. In 1896, a major reservoir was built across the once-thriving Arkavathi River north-west of the city: Hessarghatta lake. Its water was filtered naturally as it moved through the rural channels to the centre of the city. At the time, Bangalore had about 180,000 people and this system promised to deliver approximately 10 gallons per person per day; but use in the British-occupied cantonment was higher and therefore new projects were required, as well as strict rationing for the Indian population.

According to our interviews and historical accounts, the areas surrounding the city before the 1970s were richly fertile with planted forests, pastures (known in the local language of Kannada as *gomalas*), and water infrastructure that were supported, maintained and protected through overlapping political, cultural, material, and religious rules and norms, locally enforced. Some forests (*gundu thopes*) were protected shrines; some water tanks were religious artefacts; and typical village water-use practices allowed for food crops to be irrigated from the lakes only when the rains were insufficient, thus maintaining the recharge capacities of the water tanks and lakes built in their area (Mundoli et al., 2018; Mundoli, Manjunatha, & Nagendra, 2016). Planted crops were commonly low-water consumers.

In the villages we studied to the north of the city (incorporated as part of the expanded 'global city' as of 2007), villagers repeatedly stressed their dependence on, and the vitality of, the networked lakes, a watershed necessary to generate water for rural and urban needs. Today, they argue, the watershed and catchment have been damaged, and many of the lakes and channels built over with new urban and industrial structures. For example, in the eight villages displaced by the construction of the airport, completed in 2008, residents who were still in place and those resettled stated how they once shared the use and maintenance of their local lake, Bettakote (interviews in 2015, 2016 and 2017). Many of the lakes had been linked by channels across the region, and water overflow filled lakes on a lower gradient, and water that flowed over lake banks allowed for an extra crop seeding for local farmers.

H. S. Sudhira, a scientist who has mapped the region's old and new water infrastructures, suggests that the system was dense and intricate (Figure 1). Sudhira has used mapping technologies to find old water channels buried beneath existing buildings, including government ones. In our interviews with him, he explained that it was not monsoon rains alone that allowed for human settlement across the region; rather, large water tanks were carefully crafted to store water and were a vital 'commons' that relied on community management. The vast undulating terrain around the city, when contoured by labourers, functioned as a water catchment system with water flows cascading from one water body to another, and contained water percolating down, replenishing the aquifers below (Mathur & Da Cunha, 2006).

Sudhira stresses the damaging consequences of destroying this system:

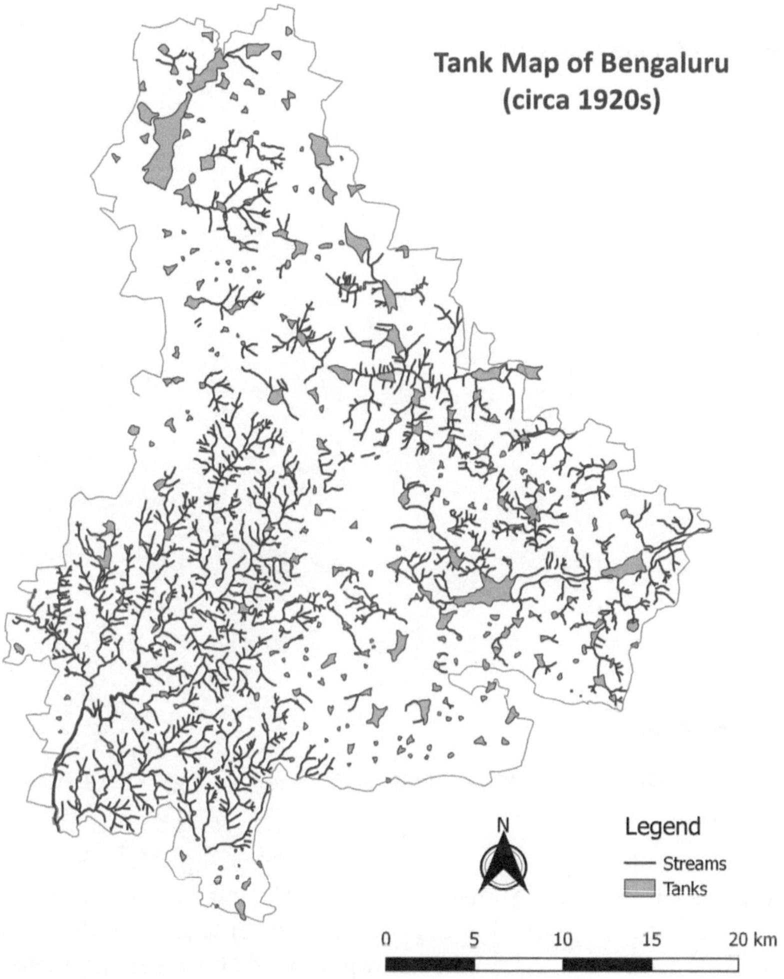

Figure 1. Map of the early 20th century dense network of tanks and canals.
Courtesy: Sudhira (2018).

> Once we started to concretize (i.e., pour concrete over) the countryside (and) turn our backs on the catchment system, in all its complex and fragile components, in a short time we began to destroy the carrying capacity of the city. (interview, June 2018)

By building over and eroding the catchment system:

> the farming communities suffered the most. But it affects us all. [...] Just look at the maps (showing us a series of time-lapsed maps), and you can know why when it rains in the city now, it floods instantly. There is no place for the rain to go. We have completely forgotten how this city was built. (interview, June 2018)

In a survey of data from the past four decades, leading ecologists found that the built-up (and concretized) area of Bangalore increased by 584% and the vegetation correspondingly declined by 66% (Ramachandra, Aithal, & Sanna, 2012). At the Indian Institute for Science, Ramachandra and his team conducted research on Bangalore's water

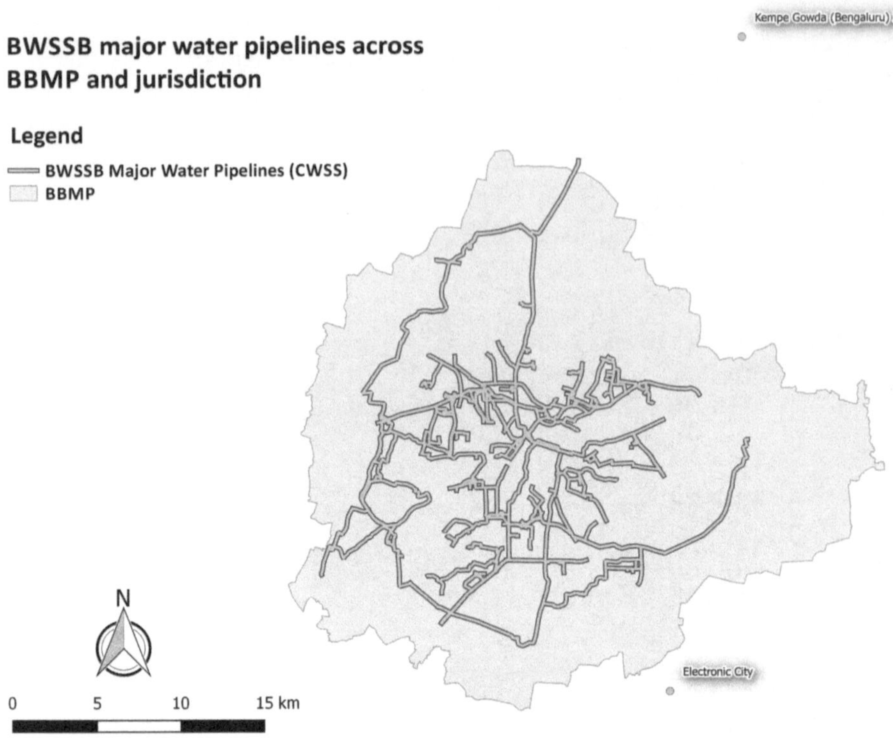

Figure 2. Map of the major Cauvery water supply system.
Courtesy: Sudhira (2018).

bodies, within the city boundaries, from 1973 to 2015, and found that 80% of the 105 water bodies have been encroached upon, converted from lake to dry bed to urban real estate, with 'lake catchments […] used as dumping yards for either municipal waste or building debris' (Ramachandra & Aithal, 2016b). While many have disappeared and converted into prime real estate, the remaining channels, or *rajakaluves*, are extremely polluted, transporting toxic sludge from north to south through the city.

A resident living alongside Varthur Lake to the south of the city started an NGO to protect the lake, one of the region's largest, from its slow death. He demonstrated how the water that ran from north to south through Bangalore drains into the second largest lake in the region (at 440 acres with a catchment area of 230 km^2), Varthur Lake, on the city's outskirts, as well as to the region's largest lake linked to it, Bellandur (the region's largest at 9000 acres). He explained:

> Only 25 years back, we would bathe, fish, and drink from the lake. The waste water that flowed from the center of the city would take more than two days to wind its way into Varthur lake; in that time, it would be filtered along the way, a natural process, through the city's and village channel system running north to south, without any major disturbance to the quality of the lake. (interview, July 2018)

Varthur has become the dump site for an estimated 40% of the urban sewage water from the city, and since 2015, it has caught fire and frothed over with industrial contaminants and sewage waste repeatedly. Ecologists based at the ATREE Research

Centre found that the only water that does flow out of the city through the once-robust water infrastructure southward is wastewater, much to the dismay of farmers dependent on this water to irrigate their fields (Thomas, Lele, Srinivasan, & Jamwal, 2017). As we show below, the water-use habits developed around the catchment systems were gradually replaced with the introduction of the piped-in Cauvery water system, and the unregulated technology of bore wells, both of which changed the dynamics of water access and generation for the city.

The hydraulic regime (II)

Expanding the city

The second hydrosocial regime is defined by public infrastructure constituted of piped water imported from the Cauvery River approximately 100 km from Bangalore. Here we look at the history, contestations and disparities associated with this system of water distribution. The new water system represented a major shift in the way water was governed and distributed, closely tied to a new model of urban growth driven by the expansion of public sector industries and based on expensive and burdensome loans from international finance institutions such as the World Bank. The new centralized governance structure changed the logic of water distribution and infrastructure management and created highly uneven social and ecological consequences.

In the period between the 1940s and the 1960s, post-independence Bangalore emerged as a hub of government-funded industries in, among other fields, radar, aeronautics, telephone and electronics, establishing Bangalore as India's 'Science City'. This new phase of urban expansion put pressure on the existing water infrastructure such that 'by the 1960s, a public debate was raging around the need for alternative water options' as Bangalore's water demands surpassed what the region's water bodies could provide (Ranganathan, 2010, p. 43). During this period, the local system of networked tanks and wells, and the two major reservoirs (Hesaraghatta and Tippagondanahalli) with their Arkavathi river base, were replaced by a second water infrastructure, which defines this *hydraulic regime.*

At the time, there was much discussion about the substantial cost of pumping water uphill to Bangalore's plateau. The main obstacle for the government was financing. The World Bank stepped in and offered a solution that had a dramatic effect on the regional ecology, the city's possibilities of expansion and also the region's governance structure. In 1964, the World Bank agreed to lend large amounts of capital over a series of loans on the condition that the State of Karnataka empower an 'independent' parastatal agency to oversee this water infrastructure. The municipal water utility, Bangalore Water Supply and Sewerage Board (BWSSB), would become accountable only to the international finance institutions that issued the loans and to the state's chief minister (Heitzman, 2004). The World Bank saw this as the most propitious way to create a powerful state agency that stood apart from local politics and the so-called backwardness and provinciality that it blamed for India's stalled progress. In fact, the World Bank had created one of the most powerful city-building agencies with little accountability to the public (Ranganathan, Kamath, & Baindur, 2009).

In the Cauvery Water Supply Project, water became a highly extractive, centralized and financially indebted enterprise. It was ruled by two forces: a parastatal agency (BWSSB) detached from public participation and oversight, beholden to the international finance institutions and the obligations of a loan contract; and the chief minister of the State of Karnataka, a political actor whose party elections are often dependent on promises of cheap resources to the voting public and construction contracts to select elites. Under this centralized-extractive regime of governance, water became an asset used to generate revenues to help pay back a loan, and its perceived scarcity conjured a dependency upon new rounds of lending by the international banks, and new political promises by the state's political leaders.

Consequently, the BWSSB became the most indebted and yet one of the most powerful parastatal agencies in Bangalore, owing more than US$35 billion to international finance agencies in 2017 alone, and yet a key agency deciding which of the new residential and industrial developments would get the crucial asset of water, an important input for value creation for investors. Reports reveal that the new loans negotiated in January 2018 with the Japanese government for a fifth stage of this project are valued at an additional US$1 billion with undisclosed interest rates and loan costs (Express News Service, 2018; Rath, 2017). Over the decades, the BWSSB has become increasingly focused on ways to manage its financial debt and to attain cost recovery from water fees rather than produce an equitable and sustainable city water distribution system.

Leo Saldanha, the director of one of the most prominent environmental organizations in Bangalore, the Environmental Support Group (ESG), and a seasoned researcher since the 1980s, tells the story of the Cauvery water system by foregrounding the politics of distribution and allocation, suggesting that power relations were embedded in the very anatomy of this water infrastructure. First, it prioritized the city of Bangalore over the rest of Karnataka, and within the city, the BWSSB prioritized the upper- and middle-class neighbourhoods. This reflected a regional political shift to the needs of the metropolitan region, and only certain segments of it, at the expense of the countryside where the bulk of Karnataka's population lives. He explains:

> [With the] favoring [of] metropolitan areas [...] farmers tended to be the losers. Farmers in 2008 actually tried to block water to Bangalore. It was a huge protest. There were (similar) riots in the 90s. The State aggregates water as a resource and you can see the disparities in the dynamics of water sharing. Today less than one half of Bangalore's water demands are met by the Cauvery. A lot of the water is allocated to middle- and upper-class areas, which means half of the city's population is tapping groundwater to survive, yet groundwater levels have fallen so sharply that many can't live off it. In one neighborhood, after a lot of government petitioning (and no results), (residents) used dynamite to blow a hole and put in a network of pipes to divert the water. (interview, July 2016)

Today the city has a population of over 10 million and the public water utility, the BWSSB, pumps around 1500 MLD (million litres of water per day) from the Cauvery River (Rajashekhar, 2015). It is distributed unevenly across the city and comes nowhere close to meeting the current water demand. By one estimate, if averaged across the population, the city's residents would have access to 75 litres per capita per day (LPCD), which falls short of the 150–200 LPCD regarded as an international standard for a metropolitan city such as Bangalore (Raj, 2013a, 2013b). But of course, the water is not evenly distributed, with Karnataka officials promising and delivering dedicated pipelines exclusively to industrial estates including the new airport and the IT corridor. As well, the city boundaries have so rapidly expanded that their water infrastructure has not been able to keep up with the sprawling periphery, despite periodical augmentation in water from the river. The spatial unevenness results in middle-class and elite households in the central areas receiving the majority of the limited water connections, while poorer communities and newly incorporated ex-rural communities rely on shared hand pumps or private water markets (Merchant, Mohan Kumar, Ravindra, Vyas, & Manohar, 2014; Ranganathan et al., 2009). In 2007, the municipal boundaries of the city were redrawn such that the BWSSB's responsibilities grew by 333% to include eight more municipalities and 110 villages, without scaling up funds and capacity, leading to a desperate lack of institutional infrastructure (Zaerpoor, 2012). Between 2011 and 2013, the BWSSB increased its withdrawal from the Cauvery from 900 to 1400 MLD. However, during this same period, its administrative boundaries excluded most of the newer peripheral neighbourhoods of the city, yet added Electronic City to the east and the airport complex with its residential complexes and Aerospace special economic zone (SEZ) to the north.

A senior administrative officer overseeing a major portfolio for the Karnataka government, one who publicly claims he is not corrupt (purportedly installing a closed-circuit television (CCTV) camera in his office), spoke of this special treatment:

> We meet every day talking about the crisis of water, at the highest levels. But all the activity around water is happening below us, and we do nothing to stop it. For example, some areas get dedicated pipelines directly from the Cauvery, while many parts of the city and rural go without. (interview, June 2015)

This official was referring to the dedicated Cauvery pipeline to the airport and the exclusive line to the IT corridor. The largest developers are making demands that they too need a dedicated pipeline to the Cauvery system in order to ensure viability for their future large-scale residential projects. He further expanded on the politics of water allocation: 'There is displacement by land, but this is displacement by water. With farmers losing access to water, and other interests demanding theirs, water has become a serious political issue.'

As current and retired administrators have acknowledged to us, the focus of the Cauvery water system has been to raise capital and expand the piped system, but based on a cost-recovery logic that privileges large consumers and large loans, not farmers or the majority of city dwellers off the public water grid. As the city managers and developers bet on an increased flow of Cauvery water – against scientific evidence that the river's volume is decreasing and the flow is less stable – they shift their priorities away from catchment water governance practices. No longer basing calculations on the *limits* of the water catchment and lakes infrastructure, instead it is based on the *potential future revenue* from higher volumes of water, which is a starkly different, risky and highly speculative calculus (Nair, 2005).

Economist Sharadini Rath explains the problem this way:

> All loans to the State are hidden in the parastatals (such as BWSSB, the manager of the Cauvery project), and they are terrifying. They are like those futuristic monsters in the movies that keep growing and take over the city! BWSSB's debt is so huge, it's crazy. From

my study, it is clear that the city and State governments only have land to sell, in hopes of paying off some of its debts. What a way to run a city. (interview, June 2016)

Speculative urban regime (III)

Land and water as financial assets

The third regime is defined by growing dependence on groundwater and the burgeoning of a private water market serviced by the proliferation of trucks (locally called water tankers) transporting water into and around the city. This regime is characterized by rapidly escalating real estate value and large-scale commercial and industrial projects that are unviable from economic and ecological standpoints. In this section, we highlight the unregulated, fragmented and crisis-oriented water infrastructures and their intersection with speculative urban dynamics.

This regime reflects the post liberalization (1991–present) era of city growth and its entanglements with the imperatives of global-city making. Bangalore, during this period, entered a period of boom, directly triggered by the growth of the IT industry and its promises of a new urban cosmopolitanism based on the high-consumptive lifestyles of the fast-growing professional class (Upadhya, 2016), with its fresh appetite for investing in portfolios of commercial and residential assets. This growth period has made tremendous demands on the city's already strained water infrastructure and pushed the city toward crisis. Bangalore has lost 79% of its water bodies in four decades and its built-up area has increased from 8% to 77% in this period, with acceleration occurring in both categories since the mid-1990s. Since 1997, the water table has fallen from 10–12 to 76–91 m, with over 40% of the population relying on groundwater (Shekhar, 2018).

Overuse of groundwater and proliferation of private water markets

Much of the population is compelled to find alternative, informal means of meeting their water needs. Individual households, depending on their economic abilities, draw upon multiple sources of water (private bore wells, Cauvery water, water trucks), while the BWSSB provides water to only some parts of the urban core, particularly older professional class neighbourhoods and business complexes. Others illegally siphon water from the Cauvery water pipelines that run directly through neighbourhoods that have not been granted access (interviews with Saldanha, June 2016, and a real estate consultant, June 2016).

Around the time when the first few phases of the Cauvery scheme were implemented (1970s), the technology for the high-speed drilling for groundwater and the installation of electric pumps (versus hand pumps) slowly became available. Over the past 15 years, this technology has proliferated and it is now common for water to be accessed directly from under one's own property with bore wells or else from private tanker operators selling groundwater pumped from farm land. The digging of bore wells and the private water market are unregulated, and there is no formal monitoring or regulation of groundwater extraction. More than half the city's water supply is met from local groundwater, from household wells and the tanker market

(Lele et al., 2013). In the rural periphery where farmland is being converted into urban real estate, farmers who are able to hold onto a sliver of land sell their groundwater beneath it to a thriving market distributed by private water tankers or trucks.

In 2015, 105,500 private bore wells were registered with the BWSSB, while over 200,000 bore wells were estimated to be unregistered (Basu, 2015). The mining department compared withdrawal and recharge and found that groundwater is being overdrawn by 378% (Basu, 2015). Across Karnataka, according to Minor Irrigation Department estimates, 5000 bore wells are being sunk *every month*; in many cases, they are replacing existing bore wells that have dried up (Kanathanda, 2017). When asked about the depth of new bore wells, farmers and urban residents stated that water tables from 300 to 400 feet have dropped in the past five to seven years to as low as 1300 feet, which is inaccessible to most people who depend on limited incomes, and at these depths often accesses toxic water.

The gap between limits of municipal water and ever-increasing demand opened up the space for a vibrant, private water market. Water tankers estimated to be in the many thousands in number emerged as an industry around the early 2000s, and have increased in numbers exponentially since 2010 (Ravishankar, 2018). A researcher who requested data from the BWSSB summarized the findings: the BWSSB had not given out any licences for commercial tankers to transport water, and it has 69 water tankers of its own. 'The remaining 3,000 to 4,000 water tankers are illegal' and 'make an annual turnover of around Rs. 1,000 crore (US$150 million)' (Deepika, 2017). Private tankers supply water to a wide range of customers, including low-income neighbourhoods, gated residential complexes, factories, hospitals, malls and hotels. The supplier market is also heterogeneous: It comprises of both small-scale, independent entrepreneurs working outside the law as well as the 'grassroots tentacles' of large-scale public and private utility companies (Ranganathan, 2014). Working-class and poor communities are dependent on neighbourhood hand pumps (which are now increasingly running dry) and the occasional water truck, while the middle- and upper-class neighbourhoods rely on private bore wells (beneath their property), partial supply of Cauvery water and the regular visits by water trucks. Late at night, one can see the city streets crowded with trucks weighed down with sloshing rural water supplies.

Although the key distribution agents are called by some a 'water mafia', this market functions as a mafia not necessarily in the sense of an economic monopoly wherein price is controlled and arbitrarily increased, but in the political sense of securing loyalty and reinforcing vote banks for politicians. While a set of actors do indeed dominate the distribution of water, this market also works as a tool for political parties and individuals to gain votes and reinforce electoral support. A senior bureaucrat overseeing water projects explained:

> The water tanker economy is symbolic of changing political power. It used to be that politicians would get votes by providing public taps, and those were fairly permanent. But today, they only supply tankers [of water], which can be offered in exchange for votes and withdrawn if necessary [especially once elections are over]. (interview, July 2016)

Painted on the side of some of these cylindrical shaped trucks are the images of local MLAs (members of the legislative assembly) and other politicians who drum up political support, votes and favours through their sponsorship of the distribution of water to constituencies. Here, we see the messy blending of economic, political and ecological factors, and the entanglement of different markets (land, water, politics).

Speculative urban and water dynamics

In this hydrosocial regime, we find the prominence of the political discourse of global-city-making that promises the 'productive' dimension of rapid urban growth, that is, better jobs, 'world-class' infrastructure, upwardly mobile class consumption patterns and limitless real estate development. This focus, however, fails to take seriously the ecological and social implications of the recent shift of business logic and activity centred on large-scale land acquisition and related financial speculation. Changes in land use and the expansion of the city are driven not just by the spread of industrial expansion, built infrastructures, new residential and commercial localities *but* also by speculative strategies for making 'money from money' – the spread of strategies for making financial gains that include debt finance, land used as collateral, real estate investment trusts (REITs), initial public offerings (IPOs) and share price fluctuations, and land acquisition for speculation. These financial strategies do not necessarily lead to greater production or wiser infrastructural management.

In 2005, to support the national and Karnataka governments' sales pitch for Asia's newest global city, land protection laws changed to allow foreigners to buy land and for land brokers to convert rural land more easily into urban real estate. With these dramatic nationwide reforms, the central government made it possible for anyone to convert rural land into a financial asset without any promise or commitment to make it 'productive' for the economy, which was the official criterion for rural-to-urban land transfers before the 2005 legislation (Levien, 2018). To Bangalore's north, there was a rush to acquire tens of thousands of acres surrounding the new international airport and its promise to turn the region into the northern wing of the new global city, creating urban real estate with unlimited value potential. Rumours that aerospace and IT/biotech firms were ready to move in sparked land value increases such that Bangalore had never before seen. In the Devanahalli area, we found farmers whose land had been acquired early for 500,000 rupees per acre (approximately US$7000), which, in 2017, sold for as high as 50,000,000 rupees (US$700,000), a 9900% value increase. As one land broker explained: 'Much of the land sold by small farmers exchanged hands 8 to 10 times, with most of the land value hikes not landing in their bank accounts, but in the hands of land brokers, government agents, and developers' (interview, July 2015).

Bangalore's global-city transformation began in the late 1990s with the boom in the software sector capturing the imagination of investors locally and worldwide, and it thrust the farming community into a risky and volatile environment. Many of the small farmers, mostly low caste and Dalit, were being pressured to sell immediately for 'piles of cash, the likes of which most small farmers had never seen before', explained a Dalit farmer advocate. 'How could they ask for more, they had no idea that that same plot of land would be sold the next day for many times the price they received' (interview, July 2015).

As noted above, much of the land surrounding the city is within a protected 'green belt' of village commons, government forest and ecological terrain that supported the complex water catchment system. Thus, this era of the land rush included the acquisition of the land on which the water catchment system was built. Whereas under the previous hydrosocial regime, the catchment suffered from neglect and encroachment, starting in the mid-1990s, it was being actively and aggressively dismantled so it could be converted into an asset for the real estate industry (with its roots in the political establishment). With this land rush came the necessity to find systemic water alternatives to both the catchment system and the Cauvery, since they both proved inadequate in meeting the needs of the rapidly expanding city. Consequently, the speculative-urban hydrosocial regime is marked by a dramatic rise in private bore wells being sunk and tapped, and the rapid growth and spread of a private water market in which rural bore well water is distributed by thousands of water tankers plying the city streets, collectively sucking dry the water aquifers.

Speculative investment in real estate and the changes in land and water use are inextricably tied to the hollowing out of the agrarian economy over several decades. Agriculture no longer provides a sustainable means of livelihood for small, highly indebted farmers in the region, and the state offers much less support than in the past, except low-wage income when there is no employment (Indian Institute of Science, 2015; Sainath, 2013; Vasavi, 2009). Our research in areas surrounding the airport documents the steady decline of agricultural livelihoods that makes it easier for the real estate market to access and convert land and water resources into commodities, the latter pumped by farmers as their last, but not lasting, asset to sell. The depletion in surface water and reliance on groundwater have affected agricultural and cropping patterns in the area significantly. Residents of these villages drew our attention to the falling water table by noting that bore wells were drying up (interviews in 2014, 2015 and 2016). One farmer in Bettakote village in the Devanahalli area said, 'We know that borewells reduce groundwater. But what else do we do to live? We have to live don't we?'

As water scarcity increases, poor farmers with small tracts of land do not have sufficient resources to invest in bore wells and hence they grow dependent on wealthier farmers for work. Small farmers who have bore wells find it difficult to bear the repair and maintenance costs the bore well requires when pumping at such a deep level. We were told about many instances of bore wells going dry. An official at Bettakote Gram Panchayat (a local governing body) estimates that 175 bore wells have been dug in just this village alone over the past 10 years. (The once sprawling 300-acre Bettakote lake was drained and encroached on by the airport authority in 2008.) Many non-elite villagers in the region, including those who own some land, have now turned to extremely low paid, daily wage work. There is an evident shift in labour from agricultural to non-agricultural occupations, as water scarcity is forcing villagers to stop farming. Interviews with teachers and principals at two schools near the airport complex suggested that local students are no longer drawn to agriculture and their parents actively discourage them from farming. Farmers attribute water scarcity in the region as a primary reason for why agriculture has become untenable (interviews in July–August 2015 and 2016).

Grandiose master plans of development projects, and the financial imperative to hoard land to anticipate these future projects' expansion, remain strangely impervious to the fact that water represents a limit: neither Cauvery water nor the crippled catchment system can accommodate the global city. Similarly, the notion that private tankers can indefinitely feed the proposed industrial, residential and commercial zones is untenable (Mahapatra, Chanakya, & Ramachandra, 2011). Even though research conducted three decades ago documented the water shortage in this area (Asian Institute of Technology, 1990; Reddy, Venugopal, Reddy, Madhukeshwar, & Lingaraju, 1992), it did not prevent the government from identifying the Devanahalli region for large-scale airport development project and an array of potential future industrial and residential projects including luxury gated communities and IT campuses.

Ironically, the spike in land value has recently prevented the government from following through with land-acquisition plans for these future ambitions as it gets ensnared by the very process in which it has participated. A top-level bureaucrat in Karnataka's administration admitted:

> It's going to be difficult to develop anything close to Bangalore. Land is not available, the cost is very high. Huge shortage of land. There is also a shortage of water. That airport area is very deficient of water. If you want to purchase a thousand acres then you have to spend a thousand crore rupees ($150 million). Is it worth it? It's exorbitant. It's very easy for industry to come in when land is cheap. So the high land costs in Karnataka are causing a bit of a set back to the setting up of industry. Beyond the airport we have reserved 1000s of acres for industrial activities. (interview, June 2015)

Government agencies acquire land from farmers with no guarantees of industrial buy-in or use; the land is no longer designated as for farming only, and it loses its protective designation as 'green belt', so that fertile land and the rural water infrastructure become officially incorporated into the speculative real estate economy. In the airport region, land prices have increased greatly over the past decade, jarring the local economy and setting new priorities led by land bankers, members of government, the business community and, more recently, foreign financial institutions. Moreover, the state, despite very low interest from industry, continues to acquire land even when it cannot afford to. It goes to extreme lengths, including taking on additional debt to secure this land. For instance, in 2014, it borrowed 1000 crore rupees (US$150 million) to fund land acquisition on behalf of the IT sector (Kumar, 2014), an industry that is not looking to build new facilities and offices. These shortfalls have had a tumultuous effect on the land-based economy. For example, no companies have committed to build and move into the IT Investment Region, even while the government has been acquiring land from farmers who were displaced and paid a pittance for their land and paid nothing for their dependence on the surrounding commons land (interviews in 2014–17). Moreover, employment growth in the IT sector in Bangalore has faltered, with the sector laying off thousands of employees in 2017 and moving into less labour-intensive services (Narayan, 2017; Sridhar, 2017; Subramanian, 2017). The IT sector no longer can absorb the growing workforce, making the trend of investing in land and water for speculative purposes more prominent by comparison.

Because there is no industrial interest in the site, the price of land offered to farmers is relatively low; but because land prices are relatively low, the business of acquiring land becomes a speculative enterprise in its own right – buy and sell rather than build. A senior official candidly expressed grave concerns about the plight of the airport region and ominously mentioned 'vested interests' that link government departments with 'the real estate mafia'. In this revealing interview he cited the 'misuse of land conversion laws where land is legally converted simply to increase its value' and continued to say with deep scepticism: 'and they claim the area will develop despite water shortage and huge jumps in land value!' (interview, June 2016).

These critical voices within the upper echelons of the state's bureaucracy demonstrate the extent to which the peri-urban zone is governed by the logic of speculative urbanism (Goldman, 2010, 2011). The lure of real estate markets and land speculation overrides long- or even medium-term considerations of a land-based productive economy and/or sustainable water infrastructure. Officials at the local planning authority acknowledged to us that the water shortage was a major unresolved issue and there were no concrete plans to address it. Meanwhile, global-city plans drive land acquisition, displace communities and encourage the selling off of the ecological commons and waterscapes.

On 1 January 2014, the city government was forced to declare itself bankrupt when the public banks refused to lend any more money or manage its insurmountable debts. The banks required collateral as liability for the city's debts, and the city offered a portfolio of public buildings and land, including the town hall, a maternity hospital, central public markets, and cemeteries and slaughterhouses (Chaturvedi, 2014). Because of the risky strategy of offering tax-free and subsidized land and water, global-city infrastructural projects never earned city agencies enough money to support these speculative projects. A 2013 audit of the Bangalore City Corporation, the body that runs the city, found that monthly expenditures for basic maintenance, salaries/pensions etc. were being paid from high-interest bank loans and not property taxes and user fees – a risky and unsustainable mode of governance.

In sum, we find that water is simultaneously a problem, a crisis, a limit and an opportunity to access land from distressed farmers. Water crisis is not solely an urban or a rural phenomenon, but intersects with the policies of rural–urban disinvestment and speculation. Under the imperative of speculative urbanism, the value of land and water, as well as those who live off of them, is being heavily discounted as a strategy to further entice financial investors to invest. This hydrosocial regime exposes the paradox of water crisis. It represents an actual limit to agrarian livelihoods but does not present an absolute limit to financial speculation and government projections of urban development. The key point here is that water scarcity is a precondition for the easy conversion of agrarian land into urban real estate and yet its scarcity does not prevent speculation and grandiose projections of urban development. However, the current water crisis is so overwhelming that state officials at various levels of the bureaucracy admit in our interviews, and the media highlight, that it has become a major obstacle. This sharp contradiction of speculative promises and realities regarding acute water shortages and untenable land prices defines this last crisis-based hydrosocial regime.

Conclusions

Of materiality and liquidity

In this paper we have analyzed Bangalore's water crisis through the lens of three hydrosocial regimes, each analytical frame featuring an entangled and complex terrain of water access, water governance and urban expansion. We have paid special attention to the intersecting dynamics of urbanization, water systems and speculative financial logics. Relational and scalar dimensions reveal the growing influence of local, national and global forms of finance capital with disruptive effects on markets of urban and rural land, water and labour. Specifically, real estate developers have had no obligation to preserve and strengthen the existing precarious water infrastructure and state agencies have prioritized risky ventures into land acquisition and economic zone development and ignored protection and regeneration of water generation systems, including the crumbling catchment-based water infrastructure. Moreover, since the early 2000s, developers have aggressively built atop water infrastructure, affecting the survivability of the countryside as well as the city in which they have heavily invested. On the city's periphery, we find that developers – with the aid of the local planning authority – primarily acquire rural land to hoard for future profit or for water-intensive built-up economic zones, which have enticed private equity firms from as far as New York, eyeing the arbitrage opportunities as developers are cash strapped and their land banks (with sky-rocketing values) can be leveraged as collateral. Poorer farming communities on the urban periphery, meanwhile, have little choice but to sell their land and water resources. Land is bought and sold at great speed, with values rising at every transaction, even while supplies of water disappear and ecosystems degrade.

Since 2010, both the real estate sector and city government agencies have become indebted to foreign corporate lenders, and increasingly to Wall Street-based private equity firms (Goldman & Narayan, 2018). By 2017, with real estate firms' and developers' debts rising, public banks stopped lending to them, creating a widening market opportunity for Wall Street firms such as Goldman Sachs, Blackstone and KKR; they have used their powerful position in the market to obtain high rates of returns from their short-term investments in and around Bangalore. According to consultant company Knight Frank, by 2017, 'close to 60% of the real estate sector's institutional funding requirement' came solely from private equity firms, 'in sharp contrast to 2010 when less than one-fourth funding came through this channel' (Rathi, 2017, p. 12). In the past few years, a string of real estate deals has earned US-based private equity firms annualized rates of returns in India much higher than their returns in China and the United States itself (Alexander & Antony, 2018). Blackstone, as one example, has become India's largest office space owner since 2014, buying up depressed assets across India's major cities and their rural peripheries, earning sizeable rates of returns and fostering a new global market in tradable securities based on the future appreciation of local rents. Because of this wealth accumulation by financial firms, the city has become economically and politically dependent upon the speculative and volatile real estate market; it is a market that can only grow with relatively cheap or discounted access to land, real estate and water.

Thus, we find that crisis pervades, expands and intensifies. The destruction of the catchment-based hydrosocial regime and the untenable exploitation of groundwater laid the groundwork for the contemporary crisis. The series of events leading up to the large-scale pollution and privatization of the city's many lakes have 'impacted the resilience of the entire city' (Nagendra, 2016, p. 174). In the 1970s, when a highly centralized, capitalized and authoritatively managed water infrastructure was developed, leaders expanded the city with the promise of an abundant river water supply and the hubris to turn its back on the age-old water catchment system that built and shaped the city for generations. Today, under a speculative urbanist logic, Bangalore's tensions between hydrosocial materiality and financial liquidity have never been as profound.

Acknowledgments

The authors thank Anuradha Sajjanhar for excellent support in the research and editorial process. They are indebted to Rachel Schurman for crucial intellectual and editorial input; and Manjunath, Mallesh, Prasanth, Vinay Baindur, Vidya P. S. and Ini Periodi for providing invaluable research support. H. S. Sudhira created two excellent maps for this paper and was always available for interviews and discussion. The authors are grateful for the feedback received from scholars hosted by Wageningen University's 'Urban Water Challenges' workshop, 2016, especially the editors of this special issue – Leila Harris, Rutgerd Boelens, Lena Hommes and Gert Jan Veldwisch – who offered extremely helpful comments, their colleagues at the Institute for Social and Economic Change and National Institute for Advanced Studies in Bangalore; and Carol Upadhya, Hemangini Gupta, Vinay Gidwani, Helga Leitner, Eric Sheppard, Lalit Batra, Anant Maringanti, Leo Saldanha and many others in the field.

Disclosure statement

No potential conflict of interest was reported by the authors.

Funding

Research was funded by an American Institute of Indian Studies Senior Fellowship, Dr V. K. R. V. Rao Chair Professorship from the Institute for Social and Economic Change, University of Minnesota research grants, an International WaTERS research and training network SSHRC grant, and a collaborative research project on Speculative Urbanism funded by National Science Foundation [grant number BCS-1626437].

References

AIT Researchers. (1990). *Application of small-scale iron removal plant in rural water supply: Phase II. Bengaluru*. Asian Institute of Technology.

Alexander, G. S., & Antony, A. (2018, March 12). How Blackstone turned India into its most profitable market. *Bloomberg*. Retrieved from https://www.bloomberg.com/news/articles/2018-03-11/how-blackstone-turned-india-into-its-most-profitable-market

Anand, N. (2017). *Hydraulic city: Water and the infrastructures of citizenship in Mumbai*. Durham, NC: Duke University Press.

Bakker, K. (2010). Commons versus commodities: Political ecologies of water privatization. In R. Peet, P. Robbins, & M. Watts (Eds.), *Global Political Ecology* (pp. 345–368). London: Routledge.

Basu, S. (2015, July 4). Bengaluru water utility to launch door-to-door drive to check illegal borewells. *Down To Earth*. Retrieved from: https://www.downtoearth.org.in/news/bengaluru-water-utility-to-launch-doortodoor-drive-to-check-illegal-borewells-40731

BBC. (2018, March 06). Is india's Bangalore doomed to be the next Cape Town? BBC. Retrieved from https://www.bbc.com/news/world-asia-india-43252435

Bhasthi, D. (2017, March 1). City of burning lakes: Experts fear Bangalore will be uninhabitable by 2025. *The Guardian*. Retrieved from http://www.theguardian.com.

Boelens, R., Hoogesteger, J., Swyngedouw, E., Vos, J., & Wester, P. (2016). Hydrosocial territories: A political ecology perspective. *Water International*, *41–1*, 1–14. doi:10.1080/02508060.2016.1134898

Budds, J., Linton, J., & McDonnell, R. (2014). The hydrosocial cycle: Defining and mobilizing a relational–dialectical approach to water. *Geoforum, 57*, 167–169. doi:10.1016/j.geoforum.2013.10.008

Chaturvedi, A. (2014, January 02). Town hall as collateral. Retrieved from https://bangaloremirror.indiatimes.com/bangalore/cover-story/bbmp-sir-m-puttanna-chetty-town-hall-parking-complex-k-r-market-public-utility-building-jayanagar-shopping-complex-malleswaram-market-johnson-market-dasappa-maternity-hospital/articleshow/28298364.cms

Deepika, K. C. (2017, April 29). Where do tankers source water from? *The Hindu*. Retrieved from https://www.thehindu.com/news/cities/bangalore/where-do-tankers-source-water-from/article18311097.ece

Express News Service. (2018, January 25). JICA inks Rs 2,500 crore deal for water project in Bengaluru. *The New Indian Express*. Retrieved from http://www.newindianexpress.com/cities/bengaluru/2018/jan/25/jica-inks-rs-2500-crore-deal-for-water-project-in-bengaluru-1763081.html

Goldman, M. (2010). Speculative urbanism and the making of the next World City. *International Journal of Urban and Regional Research, 35*(3), 555–581. doi:10.1111/j.1468-2427.2010.01001.x

Goldman, M. (2011). Speculating on the next world city. In A. Roy & Ong (Eds.), *Worlding cities: Asian experiments and the art of being global* (pp. 229–258). Oxford, UK: Wiley-Blackwell.

Goldman, M., & Narayan, D. (2018). *'The Vultures are Circling': Real Estate's Tryst with Global Finance in India* (Working paper). Sociology Department, University of Minnesota.

Heitzman, J. (2004). *Network city: Planning the information society in Bangalore*. India: Oxford University Press.

Hommes., L., Boelens, R., Harris, L., & Veldwisch, G. J. (2019). *Rural–urban water struggles: Urbanizing hydrosocial territories and evolving connections, discourses and identities*. Water International. *44*(2), 81–94. doi:10.1080/02508060.2019.1583311

Indian Institute of Science. (2015). *Environmental auditing of Bangalore Wetlands (ENVIS Technical Report 72)*. Bangalore: Ramachandra, T. V., Aithal, B. H., Alakananda, B., & Supriya, G.

Kanathanda, M. (2017, January 23), High and dry: Drought-hit Karnataka seeing a rise in illegal borewells. The Times of India. Retrieved from: http://timesofindia.indiatimes.com/articleshow/56726102.cms?utm_source=contentofinterest&utm_medium=text&utm_campaign=cppst

Kumar, A. (2014, November 6). Rs 1,000 crore Hudco fund to Karnataka to develop IT region. *Times of India*. Retrieved from https://timesofindia.indiatimes.com/city/bengaluru/Rs-1000-crore-Hudco-fund-to-Karnataka-to-develop-IT-region/articleshow/45060584.cms

Lele, S., Srinivasan, V., Jamwal, P., Thomas, B., Eswar, M., & Zuhail, T. (2013). *Water management in Arkavathy Basin: A situation analysis (Environment and Development Discussion Paper No. 1)*. Bengaluru: Ashoka Trust for Research in Ecology and the Environment.

Levien, M. (2018). *Dispossession without development: Land grabs in Neoliberal India*. New York, NY: Oxford University Press.

Mahapatra, D. M., Chanakya, H. N., & Ramachandra, T. V. (2011). Assessment of treatment capabilities of Varthur lake, Bangalore, India. *International Journal of Environmental Technology and Management, 14*, 84–102. doi:10.1504/IJETM.2011.039259

Mathur, A., & Da Cunha, D. (2006). *Deccan traverses: The making of Bangalore's Terrain*. New Delhi: Rupa & Co.

Mehta, V., Goswami, R., Kemp-Benedict, E., Muddu, S., & Malghan, D. (2013). Social ecology of domestic water use in Bangalore. *Economic and Political Weekly, 15*, 40–50.

Merchant, A., Mohan Kumar, M. S., Ravindra, P. N., Vyas, P., & Manohar, U. (2014). Analytics driven water management system for Bangalore city. *Procedia Engineering, 70*, 1137–1146. doi:10.1016/j.proeng.2014.02.126

Mundoli, S., Manjunatha, B., & Nagendra, H. (2016, June 24–25). Losing the commons: Rapid changes in gunda thopes in Bangalore, (unpublished). Paper presented at NIAS Bangalore's Great Transformation workshop, Bangalore.

Mundoli, S., Manjunatha, B., & Nagendra, H. (2018). *Lakes of Bengaluru: The once living, but now endangered peri-urban commons (Working Paper)*. Bangalore: Azim Premji University.

Nagendra, H. (2016). *Nature in the city: Bengaluru in the past, present, and future*. New Delhi: Oxford University Press.

Nair, J. (2005). *The promise of the metropolis: Bangalore's twentieth century*. India: Oxford University Press.

Narayan, D. (2017). Decoding disruption: Lay-offs in the software industry. *Economic & Political Weekly*, *52*(22), 16–19.

Online, F. E. (2017, March 2). Bengaluru water crisis: Karnataka faces severe scarcity, 160 of 176 taluks declared drought-hit. *Financial Express*. Retrieved from http://www.financialexpress.com.

Prakashi, R. (2017, March 6). Looming water crisis wakes up Bengaluru MLAs from slumber. *Times of India*. Retrieved from http://www.timesofindia.indiatimes.com.

Raj, K. (2013a). Sustainable urban habitats and urban water supply: Accounting for unaccounted for water in Bangalore City, India. *Current Urban Studies*, *1–04*, 156. doi:10.4236/cus.2013.14017

Raj, K. (2013b). *Where all the water has gone? An analysis of unreliable water supply in Bangalore City*. Working Paper 307. Bangalore: The Institute for Social and Economic Change.

Rajashekhar, A. (2015). *Do private water tankers in Bangalore exhibit 'mafia-like' behavior?* (Unpublished Master's thesis). Massachusetts Institute of Technology, Massachusetts. Retrieved from http://hdl.handle.net/1721.1/99090

Ramachandra, T. V., & Aithal, B. (2016a). Bengaluru's reality: Towards unlivable status with unplanned urban trajectory. *Current Science*, *110–12*, 2207–2208.

Ramachandra, T. V., & Aithal, B. (2016b, October 24). Decaying lakes of Bengaluru and today's irrational decision makers. Retrieved from: http://wgbis.ces.iisc.ernet.in/energy/water/paper/Decaying_lakes_of_Bengaluru/index.html

Ramachandra, T. V., Aithal, B., & Sanna, D. (2012). Insights to urban dynamics through landscape spatial pattern analysis. *International Journal of Applied Earth Observation and Geoinformation*, *18*, 329–343. doi:10.1016/j.jag.2012.03.005

Ramachandra, T. V., Asulabha, K. S., Sincy, V., Bhat, S., & Aithal, B. (2016c). *Wetlands: Treasure of Bangalore*, ENVIS Technical Report 101. Bangalore, India: Energy & Wetlands Research Group, CES, IISc.

Ranganathan, M. (2010). *Fluid hegemony: A political ecology of water, market rule, and insurgence at Bangalore's frontier* (Unpublished doctoral dissertation). University of California, Berkeley. Retrieved from https://escholarship.org/uc/item/05m1k2fs#author

Ranganathan, M. (2014). 'Mafias' in the waterscape: Urban informality and everyday public authority in Bangalore. *Water Alternatives*, *7*(1), 89–105. Retrieved from www.water-alternatives.org

Ranganathan, M., Kamath, L., & Baindur, V. (2009). Piped water supply to greater Bangalore: Putting the cart before the horse? *Economic and Political Weekly*, *44*(33), 53–62. doi:10.2307/25663447

Rath, S. (2017). The city as dichotomy. *Seminar*, *694*. Retrieved from http://www.india-seminar.com/semframe.html

Rathi, V. (2017). *Analysis of institutional funding in real estate*. India: Knight Frank.

Ravishankar, R. (2018, November 29). Bengaluru's rich use maximum water. *Deccan Herald*. Retrieved from https://www.deccanherald.com/city/bengalurus-rich-use-maximum-705511.html

Reddy, M., Venugopal, T. N., Reddy, G. S., Madhukeshwar, S., & Lingaraju, Y. (1992). *Status of groundwater exploitation and methodology for recharging of aquifers in overexploited areas of Bangalore, Kolar & Tumkur districts (Report sponsored by Resource Management Group)*. Bengaluru: Drought Management Cell.

Sainath, P. (2013, May 2). Over 2000 fewer farmers every day. *The Hindu*. Retrieved from http://genecampaign.org/wp-content/uploads/2014/07/Over-2000-fewer-farmers-every-day-P.pdf

Shekhar, D. (2018, March 23), Bengaluru Has lost 79% of its water bodies in past four decades. *The Economic Times*.

Sheppard, E., Gidwani, V., Goldman, M., Leitner, H., Roy, A., & Maringanti, A. (2015). Introduction: Urban revolutions in the age of global urbanism. *Urban Studies*, *52*(11), 1947–1961. doi:10.1177/0042098015590050

Sridhar, V. (2017, June 23). A dream in decline. *Frontline Magazine*. Retrieved from https://www.frontline.in/cover-story/a-dream-in-decline/article9721030.ece

Subramanian, S. (2017, October 24). India warily eyes AI. *MIT Technology Review*. Retrieved from https://www.technologyreview.com/s/609118/india-warily-eyes-ai/

Sudhir, T. S. (2017, April 28). Dry fact: Bengaluru is paying the price for killing its water bodies. *News Laundry*. Retrieved from http://www.newslaundry.com.

Sudhira, H. S. (2018, June). Unpublished presentation, Bangalore's Water Infrastructure in perspective, NSF Speculative Urbanism meeting, NIAS, Bangalore.

Swyngedouw, E. (2009). The political economy and political ecology of the hydro-social cycle. *Journal of Contemporary Water Research & Education*, *142*(1), 56–60. doi:10.1111/j.1936-704X.2009.00054.x

Thomas, B., Lele, S., Srinivasan, V., & Jamwal, P. (2017). Rethinking resilience in urbanizing river basins. *Seminar*, *694*. Retrieved from http://www.india-seminar.com/semframe.html

Times of India. (2018, February 13). Water crisis: Is Bengaluru heading for Day Zero? *Times of India*. Retrieved from https://timesofindia.indiatimes.com/city/bengaluru/water-crisis-is-bluru-heading-for-day-zero/articleshow/62893432.cms

Upadhya, C. (2016). *Reengineering India: Work, capital, and class in an offshore economy*. India: Oxford University Press.

Upadhya, H. (2017, August 16). Bengaluru's Varthur Lake Now Spews 10-Feet-High Froth. *NDTV*. Retrieved from http://www.ndtv.com.

Vasavi, A. R. (2009). Suicides and the making of India's agrarian distress. *South African Review of Sociology*, *40*(1), 94–108. doi:10.1080/21528586.2009.10425102

Zaerpoor, Y. (2012). *Putting the citizen at the heart of water management: A study of water in Bangalore* (Unpublished master's thesis). Columbia University, New York. Retrieved from https://doi.org/10.7916/D8C53SZC

The rural–urban equity nexus of Metro Manila's water system

Philamer C. Torio, Leila M. Harris and Leonora C. Angeles

ABSTRACT

This article examines equity concerns and inherent conflicts related to rural–urban water supply allocation and use, with focus on Metro Manila. Going beyond the much-discussed difficulty farmers experience from an allocation policy prioritizing urban water requirements, it shows that inequity in raw water allocation is linked to, and further exacerbated by, inequities in urban domestic water provision. Moreover, it highlights the need for broader equity reviews, using the concept of the rural–urban water equity nexus to draw attention to key equity considerations across space and scale that otherwise might remain invisible.

Introduction

While the concept of efficiency has dominated the discourse on water resources management in recent decades, particularly given broader shifts towards neoliberal and utilitarian views of water, there has been growing emphasis on issues of water justice and equity – concepts that are emphasized by NGOs, scholars and water justice advocates (Boelens, Perreault, & Vos, 2018; Lacey, 2008, Perreault, 2014). For water allocation, a focus on productive uses, market mechanisms and economic incentives has often prioritized higher-value uses in water supply, typically shifting the allocation schema from agricultural to urban and industrial uses in varied geographical contexts (Shah & Zerriffi, 2017; Whiteley, Helen, & Perry, 2008). Though efficiency is an important policy focus, there are clear costs and risks associated with singular attention on this concept as a predominant metric. This is particularly so considering that at times it is applied in ways that fail to attend to, and may even aggravate, water-related conflicts and supply challenges (Whiteley et al., 2008). As Lacey (2008) and Perreault (2014) have argued, there is a clear need to recognize values of equity and justice to counterbalance the common focus on efficiency and markets. Greater consideration of these values offers the ability to understand and address concerns of importance for populations that are socially disadvantaged and economically marginalized (Kirjan, 2012). Echoing and amplifying this suggestion, a range of water justice advocates and environmental justice scholars have highlighted equity concerns as paramount in water governance discussions (e.g., Boelens, 2009; Harris, 2013; Harris, McKenzie, Rodina, Shah, & Wilson, 2017; Zwarteveen et al., 2017).

Contributing to this focus, this article investigates the *rural–urban water equity nexus* of the Metro Manila water system, with specific attention to periods of El Niño occurrence and long dry spells. To develop and interrogate the notion of the rural–urban water equity nexus, the analysis considers equity issues related to water allocation from the Angat Dam, a multipurpose reservoir supplying domestic water to Metro Manila residents and irrigation water to rural farmers. Other authors (e.g., Shah & Zerrifi, 2017; Tabios & David, 2004) have also discussed the difficulty rural farmers experience during extremely dry conditions, particularly given an allocation policy that favours urban consumers of Metro Manila in times characterized by these conditions. However, this article goes further by showing that inequity in raw water allocation is linked to, and further exacerbated by, inequities in drinking water provision for nearby urban Manila. The concept of the equity nexus emphasizes linkages and interdependencies across the rural–urban gradient, and ways that these issues are often recalibrated across spaces and scales, as clearly elaborated by the introductory article of Hommes, Boelens, Harris, and Veldwisch (2019).

To begin with, our article discusses varied concepts of equity in water governance to lay the foundation for the analysis. The next section reviews the existing allocation protocols for Angat Dam, highlighting the inherent bias that favours urban water supply over agricultural uses, resulting in significant precarities and revenue losses for the farmers during periods of water scarcity. Metro Manila's water provision is then examined in the contexts of pre- and post-privatization scenarios, to understand conditions of urban water inequities that have evolved during these different periods, with attention to temporal shifts that may have occurred. The concluding section calls for broader equity reviews on water allocation and use in Metro Manila, using the rural–urban water equity nexus as a conceptual framework to document and highlight interlinked equity considerations that might otherwise remain invisible.

Equity: a variegated concept

The concept of equity is intrinsic in water resource management (Wegerich, 2007; Whiteley et al., 2008), and many policies stress the need for equitable allocation of water, but the concept of equity often remains undefined and ambiguous (Syme, Nancarrow, & McCreddin, 1999; Wegerich, 2007). Kirjan (2012) and Lacey (2008) point out that the lack of sufficient attention to equity in water resource management has led to significant conflicts, disempowering many consumers – at times, even denying basic rights to water for some. Arguably, the relative under-focus on equity is shifting, with growing discourses on the human right to water and politics associated with water justice over the past few decades (e.g., Boelens, 2009; Harris et al., 2017; Perreault, 2014; Roth, Boelens, & Zwarteveen, 2005; Zwarteveen & Boelens, 2014). Nonetheless, there is a clear need for conceptual elaboration, including clarification of key approaches and definitions, as well as enriched policy analysis with an equity lens for various aspects of water governance – from shifting water supply or pricing regimes, to implementation of new technologies or governance approaches. While these themes have been explored in the literature (e.g., Harris et al., 2017; Phansalkar, 2007; Zwarteveen et al., 2017), our contribution elaborates on the concept of *rural–urban water equity nexus*, offering analytical focus to highlight the multifaceted and multiscalar ways that urban and rural water equity issues are often interlinked, interdependent and mutually imbricated. Often there are complex equity trade-offs and consequences arising from spatial and temporal shifts across the waterscape, frequently entailing shifting water access, uses and conditions between rural and urban users and spaces. Many of the other contributions to this volume (2019) similarly elaborate dynamics and trade-offs between rural and urban users and uses. For instance, the Duarte-Abadia and Boelens (2019), Bleeker and Vos (2019), and Goldman and Narayan (2019) provide similar examples of rural-to-urban water transfers, and associated inequities – often in the name of modernity, global urbanism (in the case of Bangalore), or water utopias (in the case of Spain). Also with complex trade-offs, but with different dynamics that favour white commercial farmers in the case of South Africa (Wessels, Veldwisch, Kujawa, & Delcare, 2019), or export-oriented agriculture in Peru (Damonte & Boelens, 2019), here there is a complex dynamic where large industrial farming enjoys privileged access to water, at the expense of impoverished urban households, peasants and other marginalized communities. As such, all the contributions in this issue speak in different ways to dynamics of the rural–urban equity nexus. Here, with this contribution focused on Metropolitan Manila, we document and analyze the complex ways equity is renegotiated in relation to rural–urban water dynamics, as well as how shifting uses and conditions of water across urban and rural spaces have complex implications for differentiated, yet interlinked, inequities. We explore these linkages specific to our case study, and also offer some more general conceptual and analytical insights along these lines.

To further develop some of the required conceptual building blocks, equity is generally described as the quality of being fair, reasonable, impartial, or just (Kirjan, 2012; Perreault, 2014; Sajor & Ongsakul, 2007; Syme et al., 1999; Wong & Srikantha, 2014, Zwarteveen & Boelens, 2014). In the context of social policy, equity often has two principal components, proportionality and egalitarianism. Proportionality (or vertical equity) implies that individuals derive benefits according to the degree of effort they put in, while egalitarianism (or horizontal equity) implies equal treatment for everyone (Syme et al., 1999; Wegerich, 2007). The concept of equity is also closely related to distributive justice and procedural justice, where the former might relate to fairness in allocating water or providing water services, while the latter relates to the regulatory and participatory processes that ensure everyone has a voice, is heard, and is treated fairly in terms of the processes by which decisions are made (Kirjan, 2012; Lacey, 2008; Perreault, 2014; Sajor & Ongsakul, 2007; Syme et al., 1999; Whiteley et al., 2008; Wong & Srikantha, 2014). Harris et al. (2017) emphasize the importance of using a justice lens to respond to various water concerns related to access, affordability, quality, water-related hazards, or productive uses.

As noted by Phansalkar (2007) and Wegerich (2007), equity has no universally accepted definition, and varies considerably across regions, cultures and communities. This is particularly so when we differentiate equality (everyone has the same attributes or access to goods, exposure to bads, etc.), from equity, which relies more on a notion of fairness or sense of ethics – to ensure that everyone has fair and just distribution, or process, with explicit consideration of past injustice, uneven access, and other social justice concerns (see Goff & Crow, 2014; Wutich, Brewis, York, & Stotts, 2013; Greenberg, 1981 for elaboration of the difference between equity and equality). Given that equity relies on a sense of what is ethical and right, it is necessarily influenced by historical and cultural contexts, perceptions and circumstances (Boelens, 2009; Wong & Srikantha, 2014, Zwarteveen & Boelens, 2014). To this, Fraser (2000) adds the concept of cultural justice, which relates to the recognition that all individuals are full partners in social interaction, with the capacity and right to participate on par with other individuals in a manner that is not regulated by the institutionalized patterns of cultural value. Cultural justice deals with the acknowledgement of non-official water norms, customary patterns of organization and structures, deviant water rights, and other related values. Such considerations are related to, but not the same as, issues of participatory justice and engagement in water governance. In this regard, a holistic equity approach requires representation that includes recognition of cultural norms, coupled with the concept of fairness related to water resource distribution, and also procedural issues related to participation.

In addition to contextual understandings, equity considerations are affected by a diversity of values attributed to water (Kirjan, 2012; Lacey, 2008). Aside from utilitarianism, other values such as water's symbolic, religious and lifestyle meanings are critical to discussions of equity (Whiteley et al., 2008). The equity concept also reflects temporal dimensions, as it might refer to past circumstances and values the rights of present and future generations (Kirjan, 2012; Whiteley et al., 2008). Moreover, Greenberg (1981) points out that conditions of scarcity and abundance tend to influence and generate different perceptions of fairness, even within the same community. Based on a study involving 155 undergraduates of a Midwestern-US university, Greenberg (1981) notes that allocations based on needs, as opposed to those based on equality, were perceived as fair under conditions of scarcity while the use of either criterion was perceived as fair when making allocations of abundant resources. Echoing these findings, recent work by Wutich et al. (2013) shows that the perception of distributive justice in water-rich areas relates more with equality, while that for water-scarce areas relates more with the concept of equity. Here, there is a sense that perhaps when water is less available, considerations related to equity might emerge as paramount.

Inequities related to water are directly linked to asymmetries and imbalances in socio-economic status and political power (Phansalkar, 2007; Whiteley et al., 2008). While water scarcity may be attributed to such factors as population growth, inadequate water infrastructure, poor management, and ineffective institutional arrangements, water scarcity may also be generated by socio-political processes, through exclusion, bias and discrimination (Mehta, 2006). Adding to the biophysical concepts of meteorological or hydrologic drought, 'socially produced drought' refers more centrally to the ways that drought-like conditions, or scarcity, at times occur based on a host of socio-political or institutional factors (see Mehta, 2006; Mahayni, 2013 for linked discussions of socially produced scarcity) – thus directly linking conditions of water scarcity and key socio-political inequalities in particular spaces and times. Mena-Vasconez, Boelens, and Vos (2016) also note that psychological inclination or 'mimetic desire' to become like leading models (e.g., big farmers) produces subjects who try to follow the practices (i.e., water-intensive crops, claims to larger water supplies) of these 'mirrored successful actors', resulting in another form of socially manufactured water scarcity.

Rural–urban equity nexus

Adding a novel dimension to the discussions outlined above, we highlight interlinked rural–urban inequities to reconsider water allocation and use at the regional scale, attending to trade-offs and linkages between urban inequities and broader regional dynamics. Specifically, we link allocation protocols for raw water supply from the Angat Dam to considerations of key inequities observed in the context of urban domestic water provision, under two different scenarios: before and after the privatization of the Metropolitan Manila supply system. Subscribing to Sen's (2006, 2009, 2012) view about the difficulty of defining an ideal state of justice, we do not attempt to define the ideal state of equity in this context, but rather, analyze these scenarios in order to cast a spotlight on key inequities that are renegotiated as rural–urban water transfers and rights shift and evolve (Hommes et al., 2019), with simultaneous shifts in institutions responsible for water provision, in the context of Metro Manila's privatization. Acknowledging Boelens's (2009, p. 310) notion of equity as 'location-, time-, and group-specific political constructs of fairness', we also recognize that further engagement with location-specific notions of equity would be of interest for the analysis, although we are not able to attend to it in the context of this study. Likewise, while our case study is directly connected to the other spheres of equity and social justice, we will primarily focus on the issue of distributive justice, that is, the question of socio-economic distribution and redistribution. We now turn to our examination of multi-scalar and intra-regional trade-offs, notably between rural and urban areas, as well as intra-urban inequities.

Angat Dam and its water allocation protocol

Completed in 1967, the Angat Dam is a 131-metre rock fill multipurpose reservoir in a nearby province, 58 km north-east of Metro Manila (Metropolitan Waterworks and Sewerage System [MWSS], 2012a). With a storage capacity of 850 million cubic metres (MCM) of water, the dam supplies 97% of Metro Manila's domestic water requirements, irrigates 28,000 hectares of farmland in the nearby provinces of Bulacan and Pampanga, and generates 246 megawatts of electricity (MWSS, 2012a). It also functions as a flood-control facility: water release is regulated during extreme rainfall conditions to prevent heavy flooding of low-lying communities along the Angat River.

Water for irrigation is released through four main hydroelectric turbines with a combined power output of 200 megawatts, and flows downstream to Bustos Dam and the Angat-Maasim River Irrigation System before eventually reaching the farmlands (Figure 1). Domestic water supply for Metro Manila is released through five auxiliary turbines, generating 46 megawatts of electricity in the process, and flows to Ipo Dam for distribution to the treatment plants of the two private concessionaires (MWSS, 2012a). About 97% of Metro Manila's water supply comes from Angat Dam through a north-to-south water infrastructure system, serving the needs of 14.3 million consumers in 37 cities and municipalities (MWSS, 2012b).

Water supply allocation for the dam is governed by the Memorandum of Agreement on the Angat Water Protocol, signed by the reservoir's major users and policy makers, one of which is the Metropolitan Waterworks and Sewerage System,

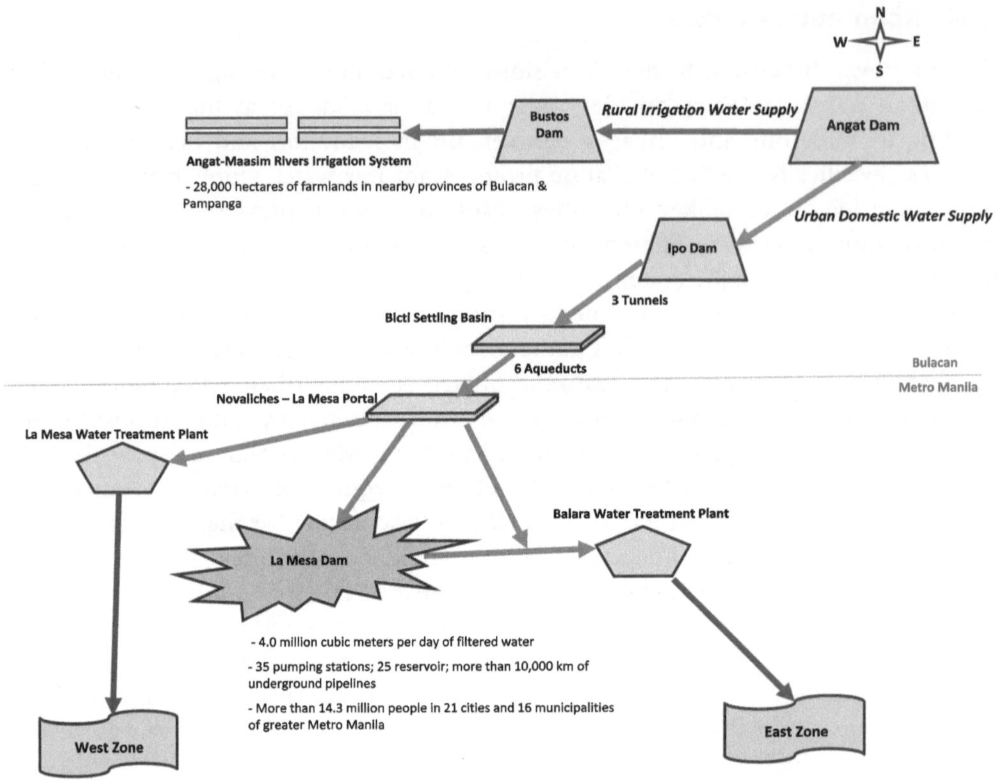

Figure 1. Rural–urban water supply system. Adapted from Torio (2016). Aside from power generation, water supply from the Angat Dam is also used for irrigation and urban domestic consumption.

the government agency responsible for water and sanitation services in Metro Manila (MWSS et al., 2013). Based on this allocation protocol, the use of water from Angat Dam is governed by the principle of 'priority in time of appropriation' for water coming from the same source – *provided that in times of emergency, municipal and domestic use shall have priority over all other uses.* Currently, this guiding principle is implemented through the Reservoir Operation Rules (Figure 2), which impose upper and lower rule curves prescribing the allocation of water among the reservoir's various users (see Shah and Zerrifi, 2017, for a more detailed discussion of Angat Dam's water supply allocation). Please note that this allocation protocol includes a miniscule allocation of 1.9 m^3/s (or only 1.4% of the total water allocation from the reservoir) for the drinking requirements of Bulacan, the province where the reservoir is located (MWSS et al., 2013). The situation that has emerged over time is one where the drinking water needs of Metro Manila are prioritized over both the productive and the drinking water needs of rural residents in Bulacan Province. What must be highlighted and examined more critically in relation to the prevailing allocation protocol are scenarios when reservoir water levels are below the lower rule curve. In line with the government's allocation policy, such scenarios, which normally occur during periods of drought, require that available water supply be

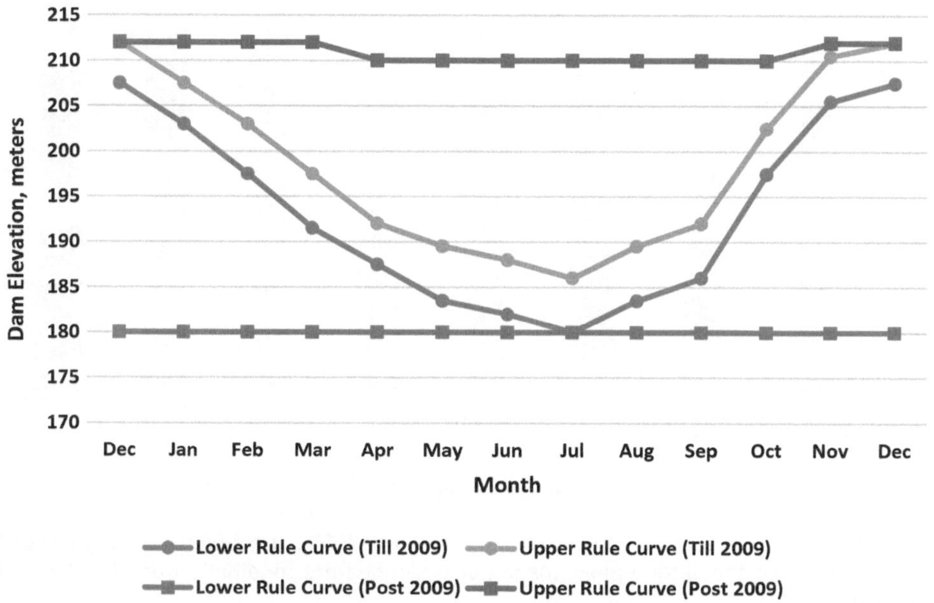

Figure 2. Operating rule curves. Adapted from MWSS (2014a, 2014b). Above the lower rule curve, all water allocations are met. Below the lower rule curve, urban water supply is prioritized over irrigation requirements.

reallocated (partly or in full) in favour of Metro Manila's urban domestic use, over the irrigation requirements of rural farmers.

Rural irrigation water supply shortfall: a frequent occurrence

From 1968 to 2010, annual water inflows into the Angat Dam were affected by the El Niño and La Niña phenomena, which are characterized by alternating periods of low and high inflows, respectively (Ortega, 2011). Of particular interest are the years of very low water inflows to the dam, resulting in water levels below the lower rule curve, which meant supply cutbacks for irrigation water. Irrigation water supply was permitted only when the water levels rose above the lower rule curve as a result of new inflows from precipitation during the wet season. A study by the Japan International Cooperation Agency, Nippon Koei Co. Ltd., and the University of Tokyo (2013) on the volume of water allocated to irrigation from 1968 to 2010 reveals several years of water supply shortfalls for irrigation with respect to the estimated mean irrigation water requirement of 600 MCM per year. Figure 3 shows these deficit periods, which generally coincided with a year of El Niño or the year immediately thereafter.

During these deficit periods, water supply for irrigation was reduced by an average of 35% from the required allocation volumes, based on approved water rights. The greatest shortfall was in 1998, when the outflow for irrigation was stopped for eight months due to a severe El Niño weather event. Cropping operations were suspended from November 1997 to June 1998, resulting in losses of 968 million Philippine Pesos (PhP) (US$24 million) (Pascua, 2007). The national irrigation agency filed a claim for compensation for these losses, but the

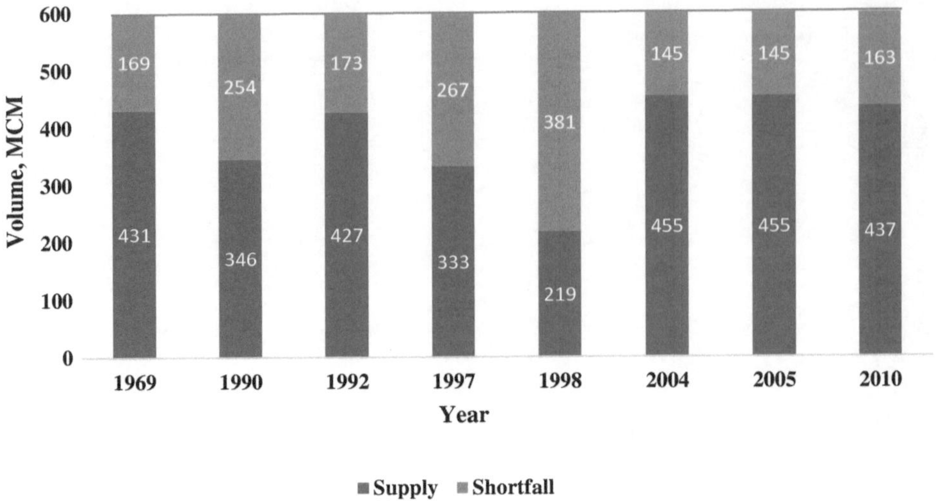

Figure 3. Irrigation water supply and shortfall. Source: Torio (2016). Irrigation water supply shortfalls normally occur during the El Niño phenomenon or prolonged dry conditions. Note: MCM = million cubic metres.

Metropolitan Waterworks and Sewerage System insisted that the reallocation was due to an unforeseen and uncontrollable event and thus was not subject to compensation under the Philippine Water Code (Pascua, 2007).

As also observed by Shah and Zerriffi (2017), the allocation protocol for Angat Dam has an inherent bias against rural irrigation water supply during times of drought, arguably at a time when the farmers would be most in need of additional water. Viewed solely through an economic lens, such a practice may be deemed justifiable, given that the opportunity cost for urban water during these conditions is estimated at PhP 5.7/m^3, while the cost of compensating the farmers for forgone revenue is only PhP 1.6–2.9/m^3 (Tabios & David, 2004). However, we agree with Tabios and David (2004), Pascua (2007), and Shah and Zerriffi (2017) that such allocation protocols result in gross inequities for the farmers, especially when they are not compensated for water supply reallocations and resultant lost revenue (which would also be invited by an equity perspective in the context of compensatory justice). Moreover, we argue that the level of inequity suffered by the farmers from an allocation scheme biased towards urban water supply is greater than has been acknowledged. We show this in the next sections by establishing the linkages to urban water inequities during pre- and post-privatization scenarios for water provision in Metropolitan Manila.

Public urban water provision until 1997

Until 1997, the Metropolitan Waterworks and Sewerage System was the provider of water and sanitation services for Metro Manila, supplying water to around 67% of the 10.9 million people in its service area, with an average supply duration of 16 hours per day (Table 1). During that time, the system's non-revenue water (NRW) level (the volume of water lost by way of leaks and pilferage) was 58%. This meant that out of a total daily water supply of 2.8 MCM, around 1.6 MCM of water was lost from the system. This proportion was worse than the average for 50 Asian water utilities, which ranged from

Table 1. Privatization scorecard.

Service indicators	1997 Before privatization	2013 After privatization*	Increase (decrease)
Population served (millions)	7.3	14.9	104%
Water supply (million L/d)	2,800	4,147	48%
Non-revenue water	58%	26%	(32%)
Water coverage**	67%	91%	24%
24-hour availability	67%	99%	32%
Water pressure (7 psi)		100%	
Water quality		100%	

*As a percentage of population served

**Post-privatization data are based on the weighted average results of the operational performance of the two private concessionaires.

Adapted from Torio (2018). Typical performance scorecard for Metro Manila's water privatization shows generally improved service levels in terms of non-revenue-water reduction, water supply availability, water pressure and water quality.

35% to 40% (McIntosh & Yñiquez, 1997). Note that water systems with high NRW require larger volumes of water supply than would be required if they were operating at higher efficiency. Moreover, the issue of NRW in the urban area links directly to the issue of water levels in the Angat Dam. Reducing NRW would have meant lower urban water supply requirements, which would have translated to higher volumes of water behind the dam. In turn, this would have meant more water for irrigation or even for setting up a buffer stock that all sectors could use during periods of drought.

Particularly in 1990, 1997 and 1998 (Figure 3), when rural agricultural water supply was cut by an average of 50%, lower NRW in the urban water system could have meant significantly smaller water supply shortfalls, significantly reducing the losses incurred by farmers during those periods. Linking these dimensions through the perspective of the rural–urban equity nexus, we see that inefficiencies in urban water provision, particularly infrastructural problems such as leaks, are directly linked to aggravated inequities experienced by the farmers. Even arguing from a narrow economic lens by considering the value of urban water supply, which Tabios and David (2004) estimate to be at PhP 5.7/m^3 (as opposed to PhP 1.6—2.9/m^3 for irrigation water), this scenario had significant cost consequences, given the large volume of water loss experienced in Metro Manila during those times. In this scenario, we point out, allocation rules, system inefficiencies and quality of urban infrastructure all recalibrate the rural–urban water equity nexus in complex and important ways. Thus, the issue of inequity is not simply about privileging urban users over rural uses: inequities are linked and dynamically reconfigured in myriad ways along the rural–urban divide.

Private-sector provision after 1997

Moving forward two decades from Metro Manila's water privatization in 1997, we ask whether this programme has caused corresponding shifts in rural–urban water equity connections. Many have already highlighted key inequities often associated with tariff increases and other changes that commonly accompany privatization, in addition to broad concerns related to the ethics of profiteering from the delivery of needed basic services, such as water (Bakker, 2007, 2010; Budds & McGranahan, 2003; Castro, 2007; Hall & Lobina, 2007; Harris & Roa-Garcia, 2013; Swyngedouw, 2005). However, our evaluation of equity issues related to Metro Manila's water privatization reveals additional complicated and ambiguous outcomes. Among other documented shifts, the private concessionaires have managed to increase service coverage to 91% in the metropolitan area, and have reduced NRW to 26% (Table 1). Likewise, in areas where infrastructural networks are in place, the private concessionaires have been able to supply high pressure and good-quality water on nearly a 24-hour basis. These major shifts hold significant potential to reduce inequities in the linked system (as indicated above), while contributing to persistent or even aggravated inequities in this context.

Conditions of inequity related to Metro Manila's urban domestic water provision generally manifest under scenarios where water remains unaffordable for those who are connected and access remains difficult for those who are not. In both scenarios, the urban poor suffer the most. Torio's (2016, 2018) research on the Metro Manila water privatization provides insights on these lingering equity concerns for poor households under varying conditions of access and affordability. In brief, low-income households in areas not covered by the private concessionaires' networks (mostly in the city's southern peripheries) consume less than the minimum World Health Organization standard of 50 litres/capita/day for basic health needs (Howard & Bartram, 2003) but pay the highest price for water among all unconnected households (Torio, 2016). Until property rights issues are settled, households in informal settlements within the networked areas must rely on services provided by community-based operators (sub-contractors for the concessionaires), at times paying 10 times as much for the last phase of water delivery. For low-income households able to acquire direct service connections from the private concessionaires, the experience of nearly 24-hour supply of high-pressure, high-quality water has doubled consumption, increasing water expenditure from 6% to 11% of average household income. This expenditure level is well above the maximum affordability limit of 5% that the Asian Development Bank and the World Bank generally recommend (Fankhauser & Tepic, 2007; World Bank, 2008).

These examples make it clear that even with improved operating efficiencies, safeguards remain necessary to ensure equitable water provision for all urban consumers. Without equitable urban water provision, water flows are often in the direction of rich households able to pay the increasing cost of water, especially when concerns related to access, affordability, quality and other dimensions continue to plague poor households. Based on existing rural–urban equity linkages, under a scenario of inequitable urban water provision, the water supply will be diverted from farmers to the taps of high-income urban households. Again, even through a narrow economic lens, the urban domestic water supply with an economic value of PhP 5.7/m^3 would benefit mostly the rich urban consumers, while the farmers suffer economic losses. Thus, there is a need to insure that urban water provision is equitable, especially under conditions where the farmers are not properly and justly compensated for lost revenues.

Note that under public-sector provision, the water allocation inequity suffered by the farmers is exacerbated by high levels of waste of urban water, with high economic value. On the other hand, under private operators, the farmers' inequity is magnified if urban water provision is inequitable, as only rich urban households benefit from the use of the highly valued water. From both efficiency and equity viewpoints, both scenarios are unacceptable. Yet they often remain invisible to policy makers, and thus generally remain unaddressed or unattended to.

Conclusion: the case for broader equity reviews

In this article, we have made a case for broader equity reviews that are multi-scale and multi-actor, given that conditions of inequity manifest in different forms, with shifting linkages, over different geographical areas, scales and time frames. We suggest that Philippine policy makers re-examine the allocation protocols for Angat Dam based on a more holistic and comprehensive view of the equity issues, with particular attention to rural users, but also in ways that engage the interlinked dimensions that emerge and shift across the rural–urban waterscape (Hommes et al., 2019). Such an approach would help policy makers realize that inequities in water governance are not confined to the geographical area of Metro Manila, nor that of Bulacan, but that key resonances, linkages and shifts exist across these spaces. What occurs in Metro Manila reverberates with equity consequences in the nearby provinces, whether due to the considerable losses from the system, or unfavourable conditions of access and affordability for poor urban households. Without a comprehensive equity review to drive water supply allocation policy, it is likely that rural farmers and impoverished and underserved urban residents will continue to suffer. Many of these inequities are worsened by ongoing inefficiencies in the urban system. Thus, our analysis offers a corrective to analyses that often position equity as a counter to efficiency focus – instead, we have shown that the ongoing system inefficiencies often propagate and worsen inequity, suggesting that these priorities and goals are linked in complex and ambiguous ways. While clearly a sole focus on efficiency often precludes an equity analysis, equity dimensions may also be served by improving system efficiency.

In the long term, the stability and reliability of water supply for the rural farmers is a major policy issue for the government. To this end, Philippine policy makers must consider building a new reservoir or any related infrastructure that ensures irrigation water supply according to the farmers' water rights, as stated in their water permits. Doing so is critical to their livelihood, and all the more so given that dryer-than-average years are expected to be increasingly common in the coming decades. It is also likely that a broader equity perspective would also invite consideration of alternative livelihood activities for the farmers, rather than a sole focus on water supply (see Shah, 2015, for a more extensive discussion of this mitigation strategy for Bulacan).

Another consideration that comes to light with an equity focus is the very real concerns related to procedural justice – concerns that are critical to rural farmers, and the urban and rural poor alike. It is clear from the allocation rules, and also from analysis of the governance of Metro Manila's water system, that there are few if any opportunities for users to participate in decisions related to pricing, irrigation supply, or domestic water needs. At the same time, we acknowledge that there are other critical aspects of equity that we have barely touched on in relation to water allocation and use in Metro Manila, such as cultural justice, gender equity, intra-rural water allocation, and other equally important issues. While these issues further highlight the need for a broader equity perspective, as we have suggested, some of the precise concerns therein are beyond the scope of the current analysis. We hope that other academics and social justice advocates will continue research on these topics to help drive policy towards more equitable conditions. With the benefit of our analysis, and the concept of the rural–urban equity nexus, we should all aim to do so in ways that attend to and foreground key connections and interlinkages, rather than focusing on single issues or concerns in isolation.

Disclosure statement

No potential conflict of interest was reported by the authors.

ORCID

Philamer C. Torio http://orcid.org/0000-0002-0413-9806
Leila M. Harris http://orcid.org/0000-0002-1700-1902

References

Bakker, K. (2007). Trickle down? Private sector participation and the pro-poor water supply debate in Jakarta, Indonesia. *Geoforum*, *38*(2007), 855–868. doi:10.1016/j.geoforum.2005.11.011

Bakker, K. (2010). *Privatizing water*. Ithaca, NY: Cornell University Press.

Bleeker, S., & Vos, J. (2019). Payment for Ecosystem Services in Lima′s watersheds: Power and imaginaries in an urban-rural hydrosocial territory. *Water International*, *44*(2), 224–242. doi:10.1080/02508060.2019.1558809

Boelens, R. (2009). The politics of disciplining water rights. *Development and Change*, *40*(2), 307–331. doi:10.1111/j.1467-7660.2009.01516.x

Boelens, R., Perreault, T., & Vos, J. (Eds.). (2018). *Water justice*. Cambridge: Cambridge University Press. doi:10.1017/9781316831847.001

Budds, J., & McGranahan, G. (2003). Are the debates on water privatization missing the point? Experiences from Africa, Asia and Latin America. *Environment and Urbanization*, *15*(2003), 87–114. doi:10.1177/095624780301500222

Castro, J. E. (2007). Poverty and citizenship: Sociological perspectives on water services and public–Private participation. *Geoforum*, *38*(2007), 756–771. doi:10.1016/j.geoforum.2005.12.006

Damonte, G., & Boelens, R. (2019). Hydrosocial territories, agro-export and water scarcity: Capitalist territorial transformations and water governance in Peru's coastal valleys. *Water International*, *44*(2), 206–223. doi:10.1080/02508060.2018.1556869

Duarte-Abadia, B., & Boelens, D. (2019). Colonizing Rural Waters. The politics of hydro-territorial transformation in the Guadalhorce Valley, Málaga, Spain. *Water International*, *44*(2), 148–168. doi:10.1080/02508060.2019.1578080

Fankhauser, S., & Tepic, S. (2007). Can consumers pay for energy and water? An affordability analysis for transition countries. *Energy Policy*, *35*(2007), 1038–1049. doi:10.1016/j.enpol.2006.02.003

Fraser, N. (2000). Rethinking recognition. *New Left Review*, *3*(3), 107–120.

Goff, M., & Crow, B. (2014). What is water equity? The unfortunate consequences of a global focus on 'drinking water'. *Water International*, *39*(2), 159–171. doi:10.1080/02508060.2014.886355

Goldman, M., & Narayan, D. (2019). Water crisis through the analytic of urban transformation: An analysis of Bangalore's hydrosocial regimes. *Water International*, *44*(2), 95–114. doi:10.1080/02508060.2019.1578078

Greenberg, J. (1981). The justice of distributing scarce and abundant resources. In M. J. Lerner & S. C. Lerner (Eds.), *The justice motive in social behavior* (pp. 289–316). New York, NY: Springer. doi:10.1007/978-1-4899-0429-4_13

Hall, D., & Lobina, E. (2007). Profitability and the poor: Corporate strategies, innovation and sustainability. *Geoforum*, *38*(2007), 772–785. doi:10.1016/j.geoforum.2006.08.012

Harris, L. M. (2013). Variable histories and geographies of marketization and privatization. In L. M. Harris, J. A. Goldin, & C. Sneddon (Eds.), *Contemporary water governance in the Global South* (First ed., pp. 118–132). New York, NY: Routledge.

Harris, L. M., McKenzie, S., Rodina, L., Shah, S. H., & Wilson, N. (2017). Water justice: Key concepts, debates, and research agendas. In R. Holifield, J. Chakraborty, & G. Walker (Eds.),

The Routledge handbook of environmental justice (pp. 338–349). London: Routledge. doi:10.14288/1.0347545

Harris, L. M., & Roa-Garcia, M. C. (2013). Recent waves of water governance: Constitutional reform and resistance to Neoliberalization in Latin America (1990–2012). *Geoforum*, *50*, 20–30. doi:10.14288/1.0132709

Hommes, L., Boelens, R., Harris, L., & Veldwisch, G. J. (2019). *Rural-urban water struggles: Urbanizing hydrosocial territories and evolving connections, discourses and identities*. Water International. *44*(2), 81–94. doi:10.1080/02508060.2019.1583311

Howard, G., & Bartram, J. K. (2003). *Domestic water quantity: Service level and health*. Geneva: World Health Organization.

Japan International Cooperation Agency, Nippon Koei Co. Ltd., & the University of Tokyo. (2013). *The study of water security master plan for Metro Manila and its adjoining areas*. Tokyo: Japan International Cooperation Agency.

Kirjan, L. (2012). Principles of equity in water law. *Queensland Environmental Practice Reporter*, *18*(81), 38–55.

Lacey, J. (2008). Utilising diversity to achieve water equity. *Rural Society*, *18*(3), 244–254. doi:10.5172/rsj.351.18.3.244

Mahayni, B. (2013). Producing crisis: Hegemonic debates, mediations and representations of water scarcity. In L. Harris, J. Goldin, & C. Sneddon (Eds.), *Contemporary water governance in the global South: Scarcity, marketization and participation* (pp. 35–45). London, UK: Routledge.

McIntosh, A., & Yñiquez, C., ed. (1997). *Second water utilities data book: Asian and Pacific Region*. Mandaluyong City: Asian Development Bank.

Mehta, L. (2006). *Water and human development: Capabilities, entitlements and power*. New York: United Nations Development Programme.

Mena-Vasconez, P., Boelens, R., & Vos, J. (2016). Food or flowers? Contested transformations of community food security and water use priorities under new legal and market regimes in Ecuador's highlands. *Journal of Rural Studies*, *44*(2016), 227–238. doi:10.1016/j.jrurstud.2016.02.011

Metropolitan Waterworks and Sewerage System. (2012a). *Proposed rehabilitation, operation, and maintenance of MWSS-owned auxiliary turbines AN4 7AN5 at the Angat hydro-electric power plant through Public Private Partnership*. Quezon City: Author.

Metropolitan Waterworks and Sewerage System. (2012b). *Water security legacy plan*. Quezon City: Author.

Metropolitan Waterworks and Sewerage System. (2014a). *Angat reservoir old rule curve*. Quezon City: Author.

Metropolitan Waterworks and Sewerage System. (2014b). *New angat reservoir operation rules*. Quezon City: Author.

Metropolitan Waterworks and Sewerage System, National Irrigation Administration, National Power Corporation, Korea Water Resources Corporation, National Water Resources Board, & Power Sector Assets and Liabilities Management Corporation. (2013). *Memorandum of agreement on the Angat water protocol*. Quezon City: Author.

Ortega, D. (2011). *Vulnerability of water resource to El Niño Southern Oscillation (ENSO): The 2007–2008 La Nina and 2009–2010 El Niño impacts on Angat Dam*. Paper presented at the APEC Climate Symposium 2011, Hawaii.

Pascua, D. D. (2007). *Water allocation between irrigation and municipal use*. Paper presented at the 4th INWEPF Steering Meeting and Symposium, Thailand.

Perreault, T. (2014). What kind of governance for what kind of equity? Towards a theorization of justice in water governance. *Water International*, *39*(2), 233–245. doi:10.1080/02508060.2014.886843

Phansalkar, S. J. (2007). Water, equity and development. *International Journal of Rural Management*, *3*(1), 1–25. doi:10.1177/097300520700300101

Roth, D., Boelens, R., & Zwarveteen, M. (Eds.). (2005). *Liquid relations: Contested water rights and legal complexity*. New Bruinswick: Rutgers University Press.

Sajor, E. E., & Ongsakul, R. (2007, December 2007). Mixed land use and equity in water governance in peri-urban Bangkok. *International Journal of Urban and Regional Research, 31*(4), 782–801. doi:10.1111/j.1468-2427.2007.00752.x

Sen, A. (2006). What do we want from a theory of justice? *The Journal of Philosophy, CIII*, 215–238. 5, May 2006. doi:10.5840/jphil2006103517

Sen, A. (2009). *The idea of justice*. Cambridge: Harvard University Press.

Sen, A. (2012). Values and justice. *Journal of Economic Methodology, 19*(2), 101–108. doi:10.1080/1350178X.2012.683601

Shah, S. H. (2015). *Water variability, livelihoods, and adaptation: A case study from the Angat river basin (Philippines)*. (Unpublished master's thesis). The University of British Columbia, Vancouver. doi:10.14288/1.0166785

Shah, S. H., & Zerriffi, H. (2017). Urban water demand, climatic variation, and irrigation-water insecurity: Interactive stressors and lessons for water governance from the Angat River basin (Philippines). *Water International, 42*, 543–567. doi:10.1080/02508060.2017.1342073

Swyngedouw, E. (2005). Dispossessing H_2O: The contested terrain of water privatization. *Capitalism Nature Socialism, 16*(1), 81–98. doi:10.1080/1045575052000335384

Syme, G. J., Nancarrow, B. E., & McCreddin, J. A. (1999). Defining the components of fairness in the allocation of water to environmental and human uses. *Journal of Environmental Management, 1999*(57), 51–70. doi:10.1006/jema.1999.0282

Tabios, G. Q., & David, C. C. (2004). Competing uses of water: Cases of Angat Reservoir, Laguna Lake and groundwater systems of Batangas City and Cebu City. In A. C. Rola, H. A. Francisco, & J. P. T. Liguton (Eds.), *Winning the water war: Watersheds, water policies and water institutions* (pp. 105–132). Makati City: Philippine Institute for Development Studies.

Torio, P. C. (2016). *Water privatization in Metro Manila: Assessing the state of equitable water provision*. (Unpublished doctoral dissertation). The University of British Columbia, Vancouver. doi:10.14288/1.0300017

Torio, P. C. (2018). Leveling the playing field for Metro Manila's impoverished households. *Water Policy, 20*, 631–646. doi:10.2166/wp.2018.191

Wegerich, K. (2007). A critical review of the concept of equity to support water allocation at various scales in the Amu Darya basin. *Irrigation Drainage System, 21*(2007), 185–195. doi:10.1007/s10795-007-9035-1

Wessels, M., Veldwisch, G. J., Kujawa, K., & Delcare, B. (2019). Upsetting the apple cart? Export fruit production, water pollution and social unrest in the Elgin Valley, South Africa. *Water International, 44*(2), 188–205. doi:10.1080/02508060.2019.1586092

Whiteley, J. M., Helen, I., & Perry, R. W. (Eds.). (2008). *Water, place, & equity*. Cambridge: The MIT Press.

Wong, P., & Srikantha, H. (2014). Equity in bulk water allocation: The case of the Mahaweli Ganga Development Project in Sri Lanka. *International Journal of Water Resources Development, 30*(4), 635–648. doi:10.1080/07900627.2014.882196

World Bank. (2008). *Rural water supply in India: Willingness of households to pay for improved services and affordability*. Washington, DC: Author.

Wutich, A., Brewis, A., York, A. M., & Stotts, R. (2013). Rules, norms, and injustice: A cross-cultural study of perceptions of justice in water institutions. *Society & Natural Resources, 26*(7), 795–809. doi:10.1080/08941920.2012.723302

Zwarteveen, M., & Boelens, R. (2014). Defining, researching and struggling for water justice: Some conceptual building blocks for research and action. *Water International, 39*(2), 143–158. doi:10.1080/02508060.2014.891168

Zwarteveen, M., Kemerink-Seyoum, J. S., Kooy, M., Evers, J., Acevedo Guerrero, T., Batubara, B., … Wesselink, A. (2017). Engaging with the politics of water governance. *WIREs Water*, 4, e01245. doi:10.1002/wat2.1245

Hydrosocial territories in the context of diverse and changing ruralities: the case of Cochabamba's drinking water provision over time

Paul Hoogendam

ABSTRACT
The article describes the history of drinking water provision of the city of Cochabamba, Bolivia, as an instructive case for analyzing the dynamic reconfiguration of hydrosocial territories and rural–urban relations over time. Cochabamba's growing drinking water demand obliged the city to expand its access to water sources, involving diverse rural areas and groups of rural inhabitants. Using the concept of hydrosocial territories, the history of Cochabamba's water solutions is explored, showing the need to analyze rural actors in a differentiated way in order to understand their (changing) role in the configuration of the hydrosocial territories.

Introduction

Cochabamba, the third-largest city of Bolivia, has expanded quickly in recent decades and has had much trouble providing drinking water to its growing population. Over time, Cochabamba's drinking water company saw the need to gain access to a growing number of water sources and to build infrastructure to conduct water to the city's distribution networks. The process to expand the city's access to water was overall conflictive and full of discussions, negotiations and struggles with the inhabitants of the rural areas in which new water sources were located (Assies, 2003; Crespo Flores, 2001; Duran, Hoogendam, & Salazar, 1998; Perreault, 2008; Swyngedouw, 1995).

The study of this contentious process of expanding water control, common to the water history of many quickly growing cities (Perreault, 2008), gives insight into how control over water sources involves configurations of diverse elements (ecological, infrastructural, institutional, social, financial and cultural) into a complex whole (de Vos, Boelens, & Bustamante, 2006; Harris & Roa-García, 2013; Swyngedouw, 2009). A useful concept to analyze these complexes is that of hydrosocial territories (Boelens, 2014; Hommes, Boelens, & Maat, 2016; Hoogesteger, Boelens, & Baud, 2016; Seemann, 2016).

In this article, the concept of hydrosocial territories is used to describe and analyze the successive phases of Cochabamba's municipal water provision, with attention to the dynamic and changing urban–rural relations that form the core of the hydrosocial territories designated for the city's water provision (see the introductory paper to this special issue: Hommes, Boelens, Harris, & Veldwisch, 2019). The article takes a critical look at the diverse geographical and historical rural realities and their influence on hydrosocial relations, as well as on sustainability and equity in the exploitation of water sources for both urban and rural purposes.

The aims of the article are two. First, it underscores the usefulness of the concept of hydrosocial territories for the analysis of water supply schemes for cities with growing water demand, as it helps in visualizing the institutional, financial, technological, political and legal means that urban groups use to extract water from the rural surroundings. Second, the article draws attention to the fact that analyzing hydrosocial territories requires the study of possible differences and contradictions among multiple rural realities. The current case shows that 'the rural' around Cochabamba is far from a uniform category. Rural actors around the city changed over time, and so did their perceptions and behaviour regarding water and water projects. In particular, it scrutinizes two distinct ruralities that exist in recent history: rural communities in the valley area and rural communities in the mountains, each with different conditions, resources, production systems and cultural beliefs, which result in diverging positions on proposed water projects (Laurie, Andolina, & Radcliffe, 2002). Accordingly, the communities in these different ruralities had markedly distinct influences on the (re) configuration of hydrosocial relations over time, as did these relations on the material conditions, cultural knowledge systems and ideology of the different groups.

The article is based on my over 20 years' experience as researcher, development practitioner, and resident in the Cochabamba region. During this time, I participated in several of the events described in this article. In 1996/97, I was involved in the founding of the Departmental Federation of Irrigation Organizations (FEDECOR),[1] which became one of the main actors in the Cochabamba Water War (see the section on the 1964-2016 period). After five years of in-depth research on farmer-managed irrigation systems and their history, in 1999, I carried out a six-month study to inventory over 50 irrigation systems in the Cochabamba Central Valley. In 2017, during a four-month follow-up study, I revisited the same systems and interviewed over 30 irrigation leaders about changes in water use, water management and water governance. In the same year, in search of alternative water sources for the Cochabamba metropolitan region, I conducted research in the mountain range north of the city during which, through interviews with local leaders and participation in communal meetings, I got to know the highland peasants' perspectives on their own water needs and their opposition to the territorial water expansion claims of the city.

The structure of the article is as follows. The next section outlines the conceptual work on which the study builds, with emphasis on hydrosocial territories and the dynamic processes that define 'the rural' in relation to water rights and water management. The third section describes the general hydrospatial and socio-political setting of Cochabamba, and the fourth, the history of Cochabamba's water provision in three phases, emphasizing the characteristics of the rural actors involved in the reconfiguration of the city's expanding hydrosocial territory. The last section presents conclusions about the applicability of the concept of hydrosocial territory and the importance of analyzing the distinctive and historically dynamic features of 'the rural' to draw conclusions on the complexity of equity and justice in rural–urban hydrosocial relations.

Hydrosocial territories and dynamic rural–urban relations

For some time now, calls have been made to study regional water problems more broadly than focusing on hydrological cycles, Integrated Water Resources Management and hydraulic infrastructure, by bringing in the social and political components of water use, distribution and management (Butterworth, Warner, Moriarty, Smits, & Batchelor, 2010; Linton, 2014; Molle, Mollinga, & Meinzen-Dick, 2008). To create a conceptual framework for these kinds of analysis, scholars have developed concepts such as the hydrosocial cycle, the waterscape, hydrosocial networks and hydrosocial territories (Boelens, Hoogesteger, Swyngedouw, Vos, & Wester, 2016; Swyngedouw, 2009; van der Zaag, Bolding, & Manzungu, 2001). These different concepts coincide in relating a variety of heterogeneous entities in water analyses, including social power and structures of governance, technologies, infrastructure, political policies and water itself. Such political-ecological perspectives on water suggest a close correlation between the transformations of the hydrological cycle on the one hand, and relations of social, political, economic and cultural power on the other (Swyngedouw, 2009).

Thus, there is some recognition that water provision schemes consist of more than physical water flows and hydraulic works. They also encompass the actors, their livelihoods, and the social relations and contradictions among them (rural–urban, catchment–distribution area); and the institutions, financial resources and knowledge systems that in a dynamic and often contested way define and redefine by whom and in what ways water is captured, collected, distributed, controlled and used (Barnes & Alatout, 2012; Duarte-Abadía, Boelens, & Roa-Avendaño, 2015).

Analyses based on these conceptual building blocks invariably show the contested and political nature of water and water governance regimes (Molle et al., 2008) and raise questions about the relationship between democratic governance and water management (Swyngedouw, 2009), the connections between hydrosocial relations and power, and the issue of inequitable distribution and water justice (Perreault, 2014; Zwarteveen & Boelens, 2014). By putting people and politics at the centre of all water issues, these studies aim to open the black box of water governance and hydrosocial relations, exploring the historical and political depth of water in society (Boelens et al., 2016; Molle et al., 2008).

This article studies the history of drinking water supply from the perspective of hydrosocial territories, which Boelens et al. (2016, p. 2) define as

> the contested imaginary and socio-environmental materialization of a spatially bound multi-scalar network in which humans, water flows, ecological relations, hydraulic infrastructure, financial means, legal-administrative arrangements and cultural institutions and practices are interactively defined, aligned and mobilized through epistemological belief systems, political hierarchies and naturalizing discourses.

This concept is particularly useful in studying the complex processes of urban water provision, since they commonly involve spatial expansion towards rural areas from which to extract water, which inevitably raises issues about territory (imagined or materialized), technology and finance, authority and power over water extraction, and the configuration of hydrosocial relations that permit, sustain or contest intended water transfers to cities.

The broadness of the hydrosocial territory concept makes it well suited for analysis of different water use configurations, but complicates its application because of the quantity of issues involved: hydrological processes, water flows, hydraulic infrastructure, water distribution, financial means, institutional development and water-based livelihoods – all in the context of the historical, geographical and technical-political processes that created and transformed them (Boelens et al., 2016). To limit the extent of any hydrosocial-territory study, one must critically evaluate which components of the territory are most determinative for the creation and expansion of stakeholder networks, water distribution forms and water governance structures.

In the case of hydrosocial territories related to water provision for cities, one of the critical issues is the interaction with the actors in the rural areas where water sources are located. Several studies have shown the unequal power balance between urban and rural populations, often leading to increasing urban control over water sources, to the detriment of rural families' livelihoods (Duarte-Abadía & Boelens, 2016; Hommes & Boelens, 2017). In these studies, 'the rural' is treated as a rather homogeneous category and not subject to specific analysis. In this article, I argue that 'the rural' is neither homogeneous nor static and draw attention to the fact that the differences and contradictions between rural realities and their changes over time have important impacts on the struggle over water projects.

Especially in peri-urban spaces around quickly growing urban centres, 'the rural' changes rapidly: on the one hand, transformation of land use, reduction or disappearance of peasant farming systems, decrease of agricultural water use, and weakening of the social organization of rural producers; and on the other hand, liberation of irrigation water and contestation over the redistribution or reallocation of water rights. All these processes occur in the wider context of politics of agrarian change and transformation of the peasantry (Bebbington, 1996).

Hydrospatial and socio-political setting of Cochabamba City

Cochabamba is a hill-surrounded city in the highland centre of Bolivia, at around 2600 m above sea level, and at the intersection of three valleys (Figure 1). The name Cochabamba, meaning 'field of ponds', reflects its original hydrological characteristics: a valley bottom with a great number of springs and natural reservoirs located above an impermeable geological layer, which forces water from the northern mountain range to surface. The Rocha River crosses the city. It was once a river with numerous ponds and the main water source for drinking water, but over the last decades became a contaminated, stinking flow of untreated sewage water coming from Sacaba Village upstream (for spatial reference see Figures 2–4).

North of Cochabamba City lies an east–west mountain range. The water division is about 4100 m above sea level. South of the division, small and steep mountain rivers flow in the direction of the valley. In the higher parts, there are many natural reservoirs, some of which have been improved through engineering works. All water sources from the mountainous area are mainly used by groups of peasant irrigators. Most of their water rights go back to the 1953 agrarian reform, when water rights, together with land, were divided among peasant communities.

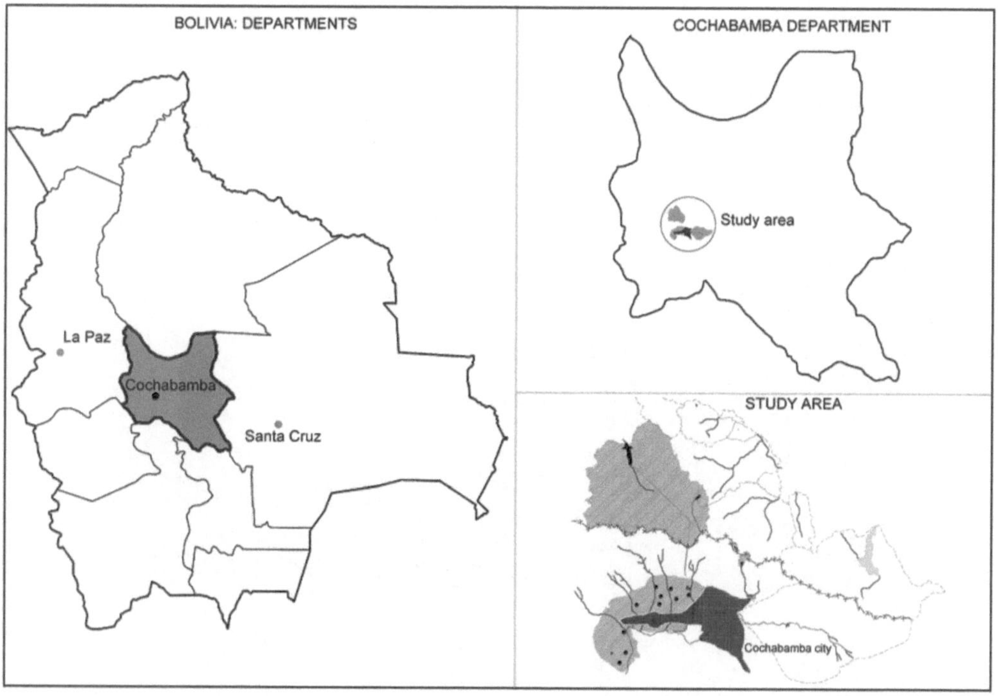

Figure 1. Cochabamba and study area in the context of Bolivia (elaboration R. Brañez).

Cochabamba City is directly neighboured by the Central and Sacaba Valleys, with a total area of around 400 km^2. The Central Valley is currently divided into five municipalities: Tiquipaya, Colcapirhua, Quillacollo, Vinto and Sipe Sipe, whereas the Sacaba Valley contains one municipality only. The distances between the municipal capitals are between 10 and 25 km. The 2016 population of Cochabamba was estimated at 850,000, and the population of the surrounding municipalities was estimated at 525,000 (Instituto Nacional de Estadística, 2017). Because of ongoing urbanization, rural areas between the urban spaces are disappearing.

The valleys surrounding Cochabamba have long been an important region for agricultural production. During the Republican period (from 1825) and up to 1953, cultivation was concentrated in *haciendas* (large land holdings), which controlled all land and water. Over time, *hacendados* (large landowners) constructed artificial reservoirs in the northern mountain range to increase water availability in the dry season (Gerbrandij & Hoogendam, 1998). After the 1953 agrarian reform, the land and irrigation systems were transferred to peasant communities, which adapted the existing systems of water rights and management to the new forms of land use and tenancy. From 1953 onwards, land and water were collective resources, with peasant families having individualized access, but over time these resources increasingly became individual and transferable assets (Gerbrandij & Hoogendam, 1998). Central Valley peasants nowadays engage mostly in the production of maize, alfalfa (for milk production), vegetables and flowers. Inheritance has fragmented land size, so that average smallholder farms nowadays are between 0.5 and 2 hectares (Hoogendam, 2017).

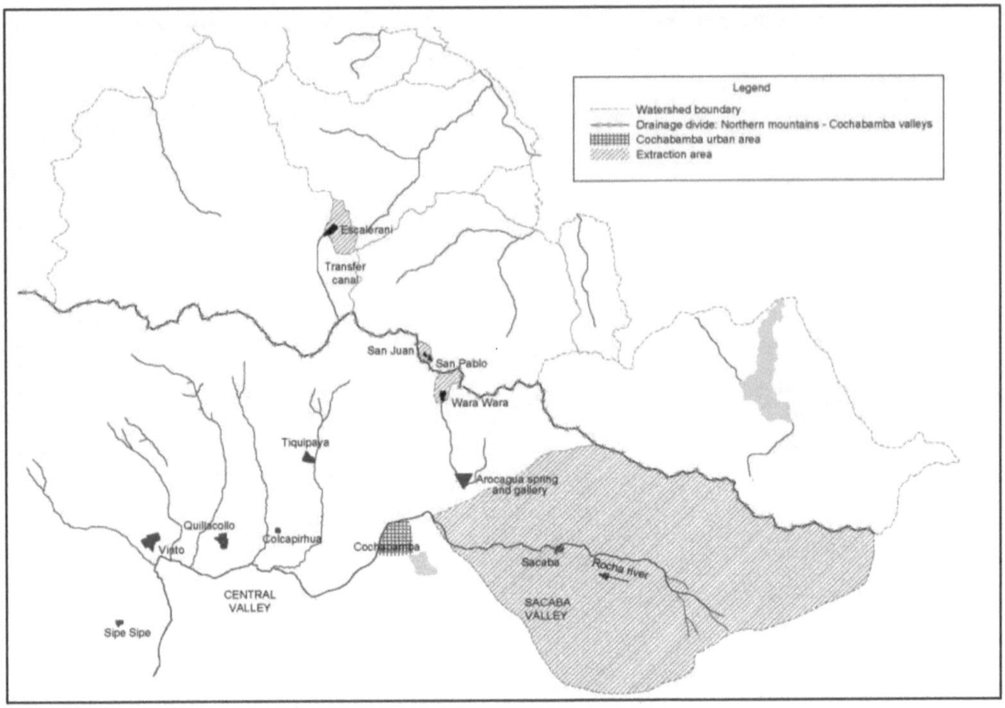

Figure 2. Extraction areas until 1967 (elaboration R. Brañez).

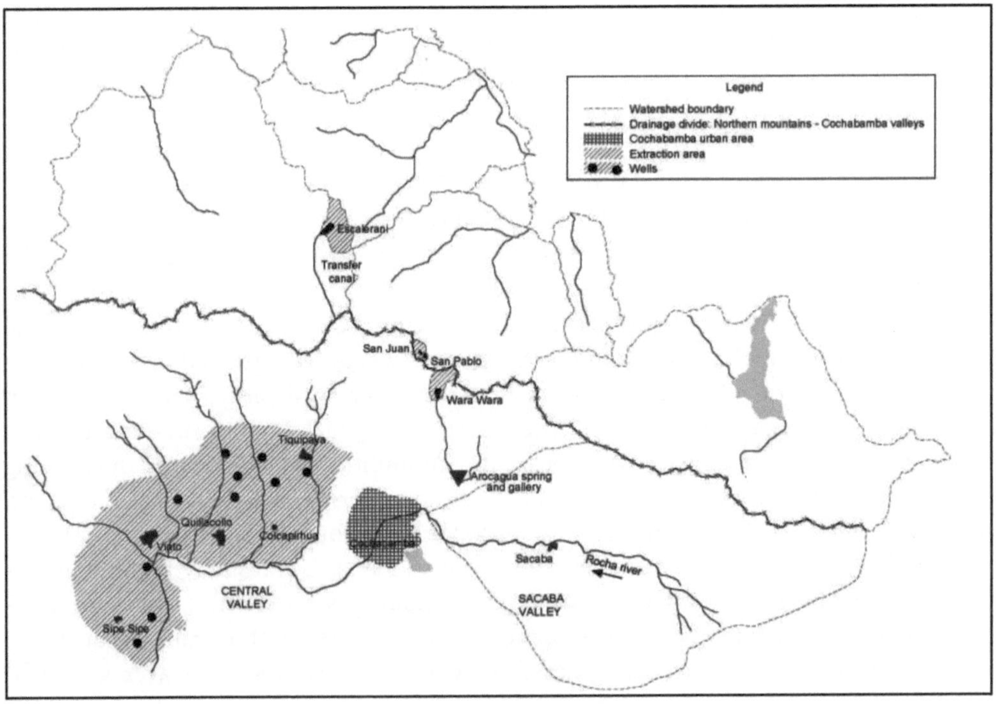

Figure 3. Extraction areas until 2016 (elaboration R. Brañez).

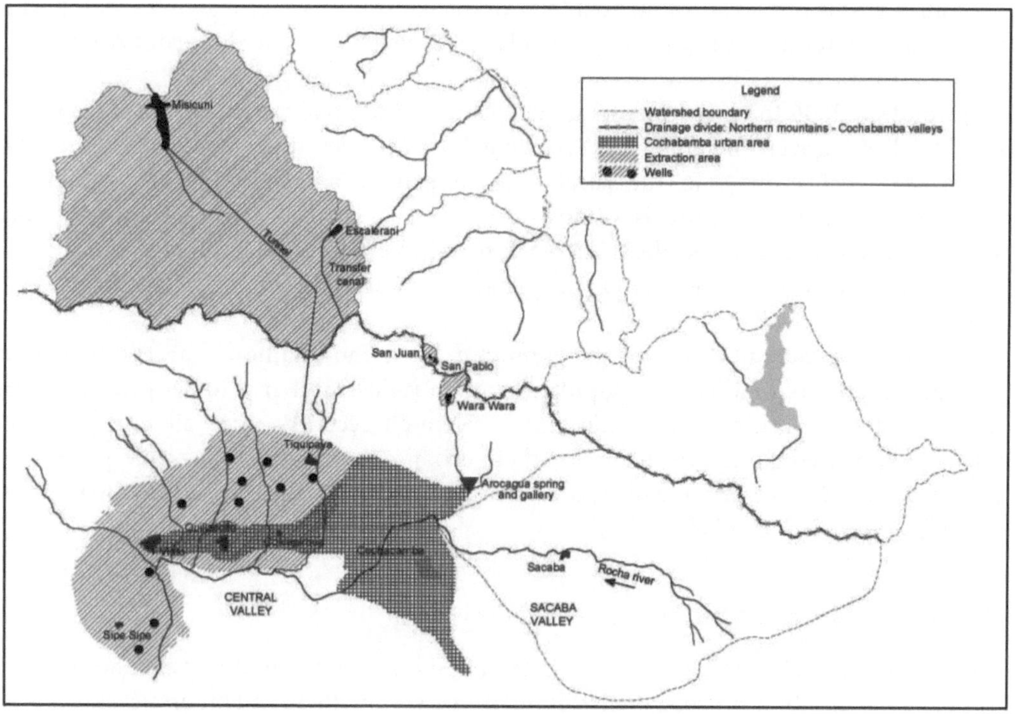

Figure 4. Extraction areas from 2017 onwards (elaboration R. Brañez).

In the mountainous rural areas north of Cochabamba, 3700–5000 m above sea level, hacienda land holdings were used for livestock farming for meat and wool production (llamas, alpacas and sheep). These haciendas hardly suffered from water problems, since precipitation levels were higher, natural reservoirs and wetlands abounded and agricultural production was limited. Here also, the 1953 agrarian reform resulted in the creation of indigenous peasant communities of former hacienda labourers, who have maintained the practice of livestock production, complemented with small plots of potato, oat and quinoa. Living conditions in the mountainous area are harsh, public services are limited, and many people live in poor conditions. In daily life and communal meetings, only Quechua (an indigenous language) is spoken. As will be shown in the next sections, the described historical differentiation between these two ruralities was essential for their position with regard to a recent water transfer project.

Besides the agrarian reform, two other main events greatly affected Cochabamba's social and political setting. First, in 1994 Bolivia initiated a process of administrative decentralization: municipalities became responsible for participative local development plans (Perreault, 2005). Until then, public investment in rural areas was realized by the Departmental Development Corporation. The new municipal responsibilities and processes of participatory planning brought, among other things, the direct involvement of local communities in defining their needs and increasing control over their livelihoods. It also led to the municipal territorialization of natural resources. One of the uneasy outcomes of this process is municipal territorial claims over water resources, hampering the implementation of water projects on a larger scale (Ministerio de Medio Ambiente y Agua, 2015).

Second, in 2006 the Movimiento al Socialismo (MAS, freely translated as Movement for Socialism) government led by President Evo Morales took office. The MAS has strong electoral support in the rural and indigenous areas. In the Department of Cochabamba, rural support is almost unanimous, especially in the mountainous areas. Over more than a decade, the MAS leadership and politics gradually transformed the collective identity of indigenous people and increased their political agency (García Linera, 2010).

In sum, Cochabamba is now neighboured by a mountainous rural area with a strongly empowered indigenous population and a valley rural area under great pressure from rapid urbanization. The latter has accelerated recently since all municipalities formally approved expansion plans for their urban areas. Over the next 20 years, it is expected that urbanization will reduce peasant agricultural land use to 30% of its current extent (Hoogendam, 2017).

Cochabamba's drinking water provision

The history of Cochabamba's water provision is similar to that of many expanding cities in the global South: it was originally founded in a spot with abundant and secure water, but its rapid population growth requires ever greater efforts to bring water from surrounding rural regions to the urban area, which commonly implies the construction of large infrastructural works and social arrangements with local actors for water extraction (Hommes & Boelens, 2017; Perreault, 2008).

In Cochabamba, the quest for water followed a similar pattern. Its history of municipal water provision can be divided into three periods. In the first period, water was extracted from the nearby Rocha River and nearby springs and mountain reservoirs. In the second period, supply was increased by drilling wells in the neighbouring valley, and, in the most recent period, water is conducted from a large, newly constructed dam in the northern mountain range. In the following sections, these three periods will be described and analyzed from a hydroterritorial perspective, with special attention to the characteristics of the rural areas involved and their influence on the water solutions.

In all periods, municipal water supply covered only part of the total water provision. At present, 56% of Cochabamba's population receives water through the municipal network (*Los Tiempos*, 4 June, 2017). It is complemented by private wells managed by individuals or water committees, and by water trucks. The hydrosocial complexity of these other sources is not the subject of this article, but has been extensively discussed by Marston (2014).

River water, spring water and nearby mountain reservoirs (until 1964)

Until the beginning of the twentieth century, Cochabamba's drinking water was mainly provided through privately owned springs and wells. For collective use, there was a small network of common taps fed from the Rocha River and surface springs in nearby

Arocagua, which were expropriated from the hacienda landowner in 1896 (Hines, 2015; Salazar, 2001). During the early twentieth century, the flow for the collective network was complemented with water from a group of shallow wells in the same Arocagua area and from an intake in a nearby mountain stream. Overall, water sources were relatively abundant; water scarcity was essentially related to conduction and distribution problems.

Near the middle of the twentieth century, the water needs of the growing urban population led to a search for mountain reservoirs to complement water provision during the dry seasons. The reason for the preference for water from the northern mountain range was twofold: first, the surface water flow within the valley (the Rocha River) had become more unpredictable and contaminated, whereas mountain reservoirs provided clean water only; and second, the surface flows decreased during the dry season, so stored water was needed to complement them.

In 1941, the departmental government responsible for the city's drinking water provision expropriated three mountain reservoirs (among them Escalerani, at 4200 m above sea level; see Figure 2) from one hacienda landowner family, and in 1961, after the municipality took over the responsibility for the city's water, other reservoirs followed (Wara Wara, San Juan and San Pablo). The negotiations over the compensation for these reservoirs were between the ruling city governors and hacienda owners, all of whom shared the same normative framework based on the recognition of private property rights over water sources (Hines, 2015). Neighbouring landowners also agreed to share part of their water in times of scarcity, relieving the pressure to expropriate their privately owned water sources, and at the same time strengthening their property rights (Hines, 2015). Figure 2 shows the different extraction areas mobilized for water supply until 1967.

Wells in the rural areas of the Central valley (1964–2016)

During the 1950s and 1960s, city governors repeatedly looked for new water sources for the quickly growing urban population, but found their search frustrated by the fierce defence of the use of the existing mountain lakes by the recently created rural communities, which had received landowners' land and water properties through the agrarian reform.

Therefore, around 1964 attention shifted towards underground sources. In its short-term plan to improve water provision, the recently created Municipal Drinking Water and Sewage Service (SEMAPA)[2] proposed to drill wells in the nearby Central Valley as the main strategy to solve Cochabamba's water problems, following recommendations of an investment plan funded by the Inter-American Development Bank.

Putting this new policy into practice, between 1970 and 1990 SEMAPA managed to drill 55 wells (120–250 m deep) in the Central Valley's rural municipalities, outside the city's jurisdiction (Figure 3). For the extraction of groundwater, SEMAPA needed only small parcels to install pumping facilities, but in exploitation it competed with dozens of shallow wells exploited by local communities, all located within a porously delimited territory. Although SEMAPA's engineers proclaimed that exploitation of lower layers would not affect existing shallow wells, it soon became evident that its groundwater extraction had a detrimental effect on existing water sources (springs, shallow wells and drilled wells) and caused severe soil depressions (Duran, Hoogendam & Salazar, 1998; FOBOMADE, 1999; Hines, 2015).

In the early 1990s, in view of opposition to the existing wells, SEMAPA proposed to drill far deeper wells (over 500 m), backed by the geological argument that removing water from a more profound aquifer would affect the water level in upper geological layers. However, alarmed by the earlier experiences, social protests by the rural municipalities arose quickly, resulting in the formation of a Water Defence Committee and a march by 10,000 rural inhabitants, villagers and peasant irrigators to Cochabamba City. These protests made SEMAPA postpone its drilling activities. After the dry spell of 1994, SEMAPA eventually drilled a deep well on a military plot, under police protection (FOBOMADE, 1999). But when after the 1997 drought SEMAPA tried to drill other wells, violent protests, including damaging of machinery and the local mayor's office, forced them to reconsider (Assies, 2003).

In a detailed account of the struggle over well drilling, Crespo Flores (2001) describes the two parties confronting each other. On the one side was the urban–governmental coalition, composed of the departmental government, Organization of Engineers, workers' union, Civic Committee, students' organization, urban neighbourhood committees, ecclesiastical leaders and the city's media, who, in summary, argued that the exploitation of groundwater was the only solution to the city's drinking water problem, that it would not affect existing water sources (as engineering reports said) and that it was legally sound, since the state holds original ownership over all water. Their discourses were based on the idea that groundwater resources were freely available, thus including the whole Central Valley in their imaginary hydrosocial territory, and that increasing water supply was the only solution to the urban water challenges. As Crespo Flores (2001, p. 56) states, 'focus on rational use and conservation [of water sources was] missing in the urban sociocultural imaginary of Cochabamba city and in the development strategies promoted by the government.'

On the other side, the Water Defence Committee, consisting of peasant irrigators and other peasant organizations, rural villagers and civic leaders, appealed for the protection of the aquifers, portraying them as a pillar of rural living and the local environment. The earlier decline of the groundwater level due to overexploitation, materially experienced in their livelihood territory, made them resistant to technical claims about non-interference and lead to questions about the ecological dimension of groundwater use, the uneven distribution of resources under market forces, and the interests of powerful urban groups (Assies, 2003; Hines, 2015).

In the context of these opposed opinions, the outcomes of the conflict were imposed by power and force, without consent. SEMAPA drilled more wells, mobilizing technology, financial means and the protection of government and public force, but only until social protests impeded further drilling through blockades and damage to equipment at the drilling sites. But, despite the protests, SEMAPA could eventually put all the drilled wells in operation, after agreements had been reached to share part of the water with the local communities.

Rise of the valley irrigators' organizations

The valley peasants' protests against SEMAPA's drilling campaign coincided with the publication of a proposed new water law, which did not recognize existing, collective customary water rights and aimed at privatization and commercialization of water (Perreault, 2005). In opposition to the proposed law, groundwater irrigators, who suffered from aquifer overexploitation partly due to SEMAPA's extraction, joined forces with river-water irrigators, who feared losing control over their collective water sources.

To make their protest more effective, Central Valley peasant irrigators founded the Departmental Federation of Irrigation Organizations, FEDECOR, mobilizing irrigation organizations from the entire Cochabamba Department. Established in October 1997, it encompassed over 100 irrigation organizations and represented over 15,000 peasant families (personal notes; Crespo & Fernández, 2001; Perreault, 2008). Several massive protest marches to Cochabamba City against the water law rapidly turned FEDECOR into an influential political actor.

The defence of the groundwater sources for irrigation was the prelude to the peasant irrigators' participation in the later Cochabamba Water War (of February and April 2000) in response to the proposed privatization of SEMAPA. The events of the Cochabamba Water War have been amply described, mainly from the perspective of drinking water governance and social movements (Assies, 2003; García Orellana, García Yapur, & Quitón Herbas, 2003; Kruse, 2005; Schultz, 2008; Spronk, 2007), diminishing the decisive role of peasant irrigators in its outcomes (an exception is Crespo & Fernández, 2001).

During the Water War two main social groups concurred in their protest: on the one hand, Cochabamba city dwellers who had suffered a sudden increase in water service fees, plus those lacking connection to the drinking water system; and on the other hand, the recently organized peasant irrigators, who protested the privatization of SEMAPA because it gave the international consortium Aguas del Tunari exclusive rights to exploit groundwater within a polygon in which many peasant wells were located. Apart from the threat of being charged for their water use by a company that had not invested in the drilling of the wells, irrigators saw it as a first step in the state's challenging of customary rights over all kinds of water sources (Assies, 2003; Crespo & Fernández, 2001; Perreault, 2008). By adapting the issue of customary rights, the protest against Aguas del Tunari 'not only … included the interests of rural irrigators and well owners, but it was also mobilizing a powerful historical, indigenous imagery concerning water' (Laurie et al., 2002, p. 266).

The urban–rural protest coalition was highly effective, combining urban marches in the city streets with irrigators' roadblocks on three interdepartmental highways. In the last phase of the struggle, the city was blocked for 10 days, which forced the government to cancel the privatization contract and to expel Aguas del Tunari from the country. Although the success of the Cochabamba Water War is usually attributed to the urban protest, it is evident that the irrigators' roadblocks on the main highways paralyzed the country and compromised the government's overall control (Spronk, 2007). Thus, the defence of customary water rights, as the cornerstone of peasants' hydrosocial territories, was one of the decisive features in the Water War – and vice versa: the outcome of the Water War strongly empowered customary water rights and was highly positive for the peasant irrigators.

An interesting aspect of the Water War was the alliance of urban and valley rural groups, who had different perspectives and interests but protested under the shared objective of blocking the privatization contract with Aguas del Tunari (Assies, 2003; Schultz, 2008; Spronk, 2007). However, this alliance was not the emergence of a new social movement (Assies, 2003; Spronk, 2007) but mainly a case of 'united resistance' (Crespo Flores, 2010). The contrasting interests within the hydrosocial territory reappeared quickly: in November 2000 peasant irrigators protested new drilling initiatives by the recently 'recovered' SEMAPA.

Decline of the valley irrigators' organizations

From 1998 to 2010, FEDECOR and the irrigation organizations were important actors in the discussions, practices and discourses about water use and water governance. The irrigation movement was even able to 'alter the institutional arrangements by which water is governed in the country: the processes of participation, the scales of decision making, and the norms by which water rights are allocated and management practices effected' (Perreault, 2008, p. 846). It was also one of the largest and best-organized groups in promoting the MAS party, eventually resulting in Evo Morales' presidency in 2006.

After this period of strength, however, the organizational importance of many valley irrigators' organizations declined, mainly because of the changing agricultural conditions in the Cochabamba Valley. Especially near the city and around the rural villages, the urbanization process weakened irrigation groups: peasant families sold their land, making irrigated land more dispersed and thus collective action for distribution and control of water flows more difficult. This resulted in the gradual abolition of irrigation in areas distant from the intake works and accumulation of water in the sectors nearby. This process of water accumulation is neither collective nor controlled, but is essentially a business opportunity for irrigation leaders, which undermines the democratic and representational character of the irrigation organizations.

In this context of weakening collective control and the accumulation of irrigation water in a few hands, an informal water market appeared. Part of the water is sold to neighbouring peasants, leading to water accumulation in fewer communities. Another part is sold to drinking water corporations, such as SEMAPA and local peri-urban cooperatives (irrigators' meeting, 1 August, 2017). Thus, under the umbrella of customary water rights, new hydrosocial relations are developing, generating water-based incomes for a small number of peasant 'water lords'.

Water transfer from the Misicuni reservoir (2016–now)

After SEMAPA was obliged to stop drilling wells, the only alternative water source left for Cochabamba was the runoff of the northern mountain range, which drains in the north-eastern direction, towards the tropical forest. To benefit from this water, it must be transferred to the Cochabamba Valley by means of tunnels or transfer canals.

One of the transfer options was the Misicuni multipurpose project, designed to provide water for drinking, irrigation and the generation of electricity. It consists of a dam in the Misicuni River (Figure 4), from which water is conducted to the Central Valley through a 19.4 km tunnel. The Misicuni catchment area has an extension of

235 km^2. It is populated by indigenous highland peasants, who mainly grow potatoes and herd their llamas and alpacas on the natural fodder that grows in the mountain wetlands.

During the Water War, all protesters, urban and rural, strongly urged the construction of the Misicuni project, since it was seen as a panacea for Cochabamba's water problems. Laurie & Marvin (1999) observe that none of the valley actors wondered or cared about the fate of the indigenous mountain population living in the Misicuni watershed and in the inundation area. As a mountain community leader stated, 'Valley people assume that nobody lives behind the mountains and that all resources are at their disposal. They do not know that we live here' (intercommunal meeting, 16 July 2017).

The Misicuni tunnel was finished in 2004, and the dam was (operationally) finished in December 2016. With the water from the Misicuni reservoir, a 120 MW hydroelectric power plant is operated, below which water is distributed for drinking water and irrigation. Misicuni water complements the previously existing water sources. The availability of the Misicuni water continues to reconfigure the hydrosocial relations and territories in multiple ways, recently especially so because valley municipalities are claiming to transfer control over the SEMAPA wells to their municipal governments to exploit them for their own population and no longer for the city.

As in many reservoir projects, a negative impact of the Misicuni dam was the displacement of over 170 peasant families, who were relocated from the valley floor to higher slopes, and the loss of a large share of their productive land. The affected families were legally compensated for their inundated land and goods, but at very low prices, due to the unwillingness of project implementers to pay prices that reflected the real replacement value of the lands and the passive and accepting role of the affected families in the negotiations (Laurie et al., 2002).

Another negative impact mentioned by mountain community leaders was the drying out of wetlands and springs due to the Misicuni tunnel (intercommunal meeting, 16 July 2017). This drying out of local water sources was an unforeseen and troubling side effect of the Misicuni project: impacts on highland water availability were not expected, and no monitoring system was put in place. Now, 15 years later, discontent prevails in the affected communities about the negative impacts of the dam and tunnel construction, and they criticize the lack of public investment to guarantee and improve their livelihoods.

Empowerment of the highland peasant communities

The current mobilization of discontent in the mountain highlands is an effect of a recent process of empowerment. Their participation and political involvement started after the 1994 administrative decentralization process and changed thoroughly after the MAS party came to power. The political agency of the indigenous rural population has increased, largely because of their active participation in both the peasant organization and the political structure of the MAS party.

The cultural and political empowerment in these peasant communities led to a shift in their concepts of territory and natural resources. At present, highland communities claim that they hold ownership of all natural resources located or originating in their territory. The right to appropriate local natural resources is now the prevailing discourse in the

Cochabamba highland area and in many other Bolivian regions with predominant indigenous and peasant populations. As the vice president stated in a water seminar, 'Today it is common that people block extraction of water from their wells or diversion of water from a river, and they even demand compensation for the use of lakes that have existed for years in their communities. This is a recent process, which started at the end of the 1990s and the beginning of this century' (García Linera, 2010).

The highland communities' empowerment and territorial discourse is giving rise to renewed discussions and negotiations over water use between highland and valley communities. Reservoirs constructed by valley dwellers are now claimed by mountain communities as part of their 'original patrimony' since they are located in their territory, which gives them the original right of use (Seemann, 2016).

Until recently these claims were hardly pressed, since the water demands of the highland communities were limited. However, in the last decade, highland communities started to suffer from less reliable rain patterns, while higher temperatures allowed them to venture into the production of marketable crops. These factors fomented a growing highland demand for water and competition over water sources. Over the last few years, mountain communities have started to tap water from rivers and springs and have taken over several mountain reservoirs (Saldías, Boelens, Wegerich, & Speelman, 2012). Some weakened valley irrigation groups have already given up their mountain water sources. Recovery claims by other groups met with violent reactions from the highland communities, resulting in injuries and one death (*Los Tiempos*: Larati reservoir, 24 June 2017; San Juan reservoir, 8 December 2016; Taquiña, Santa Rosa and Vizcachas reservoirs, 25 January 2017; Saytu Khocha reservoir, 19 November 2016). Valley irrigation leaders estimate that 25–30% of the original water from the mountain sources no longer reaches the valley (Hoogendam, 2017). Although they have had a victory in the short run, some highland peasant leaders are conscious that 'violence does not create right' and are searching for ways to establish consensus on the redistribution of water rights (personal communication, 18 July 2017).

The political empowerment in the highland areas also affects control over hydraulic infrastructure, constructed to transfer water from one catchment to another, that passes through highland communities. For decades, the transfer canals were tolerated, but because of the growing sense of territoriality, complaints are increasing. According to a hydro-electric plant manager responsible for the Escalerani canal, protests grew notably after the enactment of Bolivia's new constitution in 2007, which the president announced with a discourse about the defence and recuperation of Bolivia's natural resources and territory (personal communication, 21 July 2017). Highland communities have replicated the same discourse towards SEMAPA, threatening to cut the drinking water supply. In exchange for letting the city use water from their territory, they have demanded greater compensation for animals drowned in the transfer canal and the installation of a 20 km wire mesh fence.

Control over hydraulic infrastructure is especially relevant to the Misicuni project. The generalized discontent in the mountain communities about unjust compensation and the disappearance of their water, underlined by the imaginaries of original patrimony and territorial rights, may threaten the operation of the hydraulic works and thus undermine water provision to the city. Mountain peasants have physical control over

Table 1. SEMAPA's water sources to supply the city (SEMAPA, 2017).

Type of water source	Current flow (2012–2015)		Intended situation (2026)	
	L/s	%	L/s	%
Wells in rural valley area	453	43%	0	0%
Wells in city jurisdiction	15	1%	15	1%
Reservoirs and superficial flows	378	36%	378	29%
Misicuni	209	20%	934	70%
Total	**1,055**		**1,327**	

the area where the Misicuni reservoir is located, giving them leverage in negotiation, and the historical analysis of hydrosocial territories related to Cochabamba's water provision shows the need for sustainable hydrosocial relations to guarantee water supply in the long run. The impact of any interruption of the drinking water service is aggravated by Cochabamba's dependence on Misicuni water if most valley well water is transferred to the rural municipalities (Table 1).

Conclusions

The analysis of Cochabamba's history of drinking water supply permits conclusions on both conceptual and substantive issues. On the conceptual side, it shows the usefulness of hydrosocial territories to analyze water situations. Cochabamba's case clearly demonstrates the interactive character of multiscalar networks, which actively combine water flows, hydraulic infrastructure, financial means, legal-administrative arrangements and cultural institutions in the processes of (re)configuration of specific water outcomes. It further shows that the hydrosocial territories and relations are dynamic over time, in response to changes in any of their constituent components and resulting from contested negotiations.

As a conceptual contribution to the debates on rural–urban hydrosocial territories, the Cochabamba case shows the need to problematize the concept of 'the rural': it is not homogeneous but consists of diverse, complex and dynamic realities. In the Cochabamba region, 'valley rural' is different from 'highland rural' and is increasingly so due to ongoing urbanization and changes in hydrosocial relations within irrigation systems. The case shows the dynamic of these ruralities over time: both have suffered social, economic, political and identity changes, which profoundly affect the hydrosocial relations that underlie water provision to the city (and the local rural population). The valley peasants have successfully defended their surficial water sources and, by means of territorial arguments, are on their way to regaining control over groundwater resources. Their organizational strength put them to be in a bargaining position in relation to urban institutions and groups. In contrast, the mountain peasants have been undermined by the institutional and infrastructural assault on their territory and see 'the urban' as merely antagonistic to their own interests.

On the substantive level, analysis of the Cochabamba case shows that, contrary to other studies (e.g. Duarte-Abadía & Boelens, 2016; Hommes & Boelens, 2017), the Bolivian valley peasant population has had a considerable influence on decision making on water use and water governance in relation to the city. The valley irrigators succeeded in obtaining compensation water, defending their traditional rights, obstructing privatization, and limiting groundwater exploitation. The elements of strength of the irrigation movement were their organizational capacity, strategic alliances, enduring protests and the position of irrigator leaders in influential political positions within the overall changing power relations in the Bolivian political context.

On the other hand, the struggle of highland indigenous peasants over water is only beginning. Their claims mainly regard water sources located in their mountain territory but used by valley irrigators. The highland peasants have gradually taken over some of these sources, which is proof of the political empowerment of the highland communities, defended through the imaginary of territorial rights. Their territorial claims are putting pressure on the hydrosocial relations that until recently governed the rights of use and distribution of highland water sources, and the operation of the existing waterworks.

The latter calls attention to the notion of water rights itself. In earlier collaborative studies with other scholars, we have conceptualized water rights in the context of hydraulic property regimes, which were intrinsically related to construction and maintenance processes of irrigation works (Beccar, Boelens, & Hoogendam, 2002; Gerbrandij & Hoogendam, 1998). There is a clear need to give more attention to territoriality in the analysis of water rights, and with it, to the mechanisms of spatial control. Because of their riparian location, mountain communities can materialize their access to water, forcing former right holders to negotiate new rules on shared water management and use. This is in line with the statement of the Departmental Water Management Unit (Dirección de Gestión de Agua, 2015, p. 20) that 'today's water conflicts are larger and not precisely the result of water shortage, but also of other social causes, among them … the appropriation, concentration and commerce of water, because of notions of political, administrative territoriality (departmental, municipal and communal)'.

Finally, the analysis of diverse ruralities and their development over time demonstrates that issues of water justice need a thorough investigation of the dynamics in the rural reality. The rightful defence of customary rights during the Cochabamba Water War does not mean that future water distribution and use will remain equitable or just. In the context of urbanization, customary rights disguise processes of water accumulation in a few hands. This raises questions about the relationship between customary rights, democratic participation and the need for regulation of water and justice. In Bolivia, these issues are to be addressed in new water legislation, for which some fundamental dilemmas must be tackled. Citing Bolivia's vice president (García Linera, 2010, p. 39), 'Do we take [water] from [a group of irrigators] to give it to Cochabamba? Who has the right, who defines these rights? Can we say that 500,000 inhabitants are more important than 5,000 [irrigators]? Is it possible to make a kind of water right assemblage, understanding it as half collective and half common resource?'

Notes

1. FEDECOR: Federación Departamental Cochabambina de Organizaciones de Riego.
2. SEMAPA: Servicio Municipal de Agua Potable y Saneamiento.

Disclosure statement

No potential conflict of interest was reported by the author.

References

Assies, W. (2003). David versus Goliath in Cochabamba: Water rights, neoliberalism, and the revival of social protest in Bolivia. *Latin American Perspectives, 30*(3), 14–36. doi:10.1177/0094582X03030003003

Barnes, J., & Alatout, S. (2012). Water worlds: Introduction to the special issue of Social Studies of Science. *Social Studies of Science, 42*(4), 483–488. doi:10.1177/0306312712448524

Bebbington, A. (1996). Movements, modernization and markets; Indigenous organizations and agrarian strategies in Ecuador. In R. Peet & M. Watts (Eds.), *Liberation ecologies, environment, development, social movements* (pp. 86–109). London: Routledge.

Beccar, L., Boelens, R., & Hoogendam, P. (2002). Water rights and collective action in community irrigation. In R. Boelens & P. Hoogendam (Eds.), *Water rights and empowerment* (pp. 1–21). Assen: Van Gorcum Publishers.

Boelens, R. (2014). Cultural politics and the hydrosocial cycle: Water, power and identity in the Andean highlands. *Geoforum, 57*, 234–247. doi:10.1016/j.geoforum.2013.02.008

Boelens, R., Hoogesteger, J., Swyngedouw, E., Vos, J., & Wester, P. (2016). Hydrosocial territories: A political ecology perspective. *Water International, 41*(1), 1–14. doi:10.1080/02508060.2016.1134898

Butterworth, J., Warner, J., Moriarty, P., Smits, S., & Batchelor, C. (2010). Finding practical approaches to Integrated Water Resources Management. *Water Alternatives, 3*(1), 68–81.

Crespo, C., & Fernández, O. (2001). *Los campesinos regantes de Cochabamba en la Guerra del Agua: una experiencia de presión social y negociación* [Cochabamba's Peasant irrigators in the Water War: An experience of social pressure and negotiation]. Cochabamba: Universidad Mayor de San Simón.

Crespo Flores, C. (2001). Gestión ambiental: El conflicto por la perforación de pozos profundos en Vinto – Sipe Sipe [Environmental management: Conflict about the deep well drilling in Vinto – Sipe Sipe]. In C. Crespo Flores & R. Orellana Halkyer (Eds.), *Conflictos ambientales; dos casos: agua y territorio* (pp. 23–77). Cochabamba: CERES.

Crespo Flores, C. (2010). El movimiento nacional del agua boliviano. De la resistencia a la cooptación (2000-2007) [The Bolivian national water movement. From resistance to co-optation]. In F. Poupeau & C. González (Eds.), *Modelos de gestión del agua en los Andes, Actes y Mémoires de l'Institut Francais d'Études Andines* (pp. 111–132). Lima: Instituto Francés de Estudios Andinos.

de Vos, H., Boelens, R., & Bustamante, R. (2006). Formal law and local water control in the Andean region: A fiercely contested field. *International Journal of Water Resources Development, 22*(1), 37–48. doi:10.1080/07900620500405049

Dirección de Gestión de Agua. (2015). *Plan Director de la Cuenca del Río Rocha* [Guiding Plan for the Rocha River]. Cochabamba: Gobierno Autónomo Departamental de Cochabamba.

Duarte-Abadía, B., & Boelens, R. (2016). Disputes over territorial boundaries and diverging valuation languages: Santurban hydro-social highlands territory in Colombia. *Water International, 41*(1), 15–36. doi:10.1080/02508060.2016.1117271

Duarte-Abadía, B., Boelens, R., & Roa-Avendaño, T. (2015). Hydropower, encroachment and the re-patterning of hydrosocial territory: The case of Hidrosogamoso in Colombia. *Human Organization, 74*(3), 243–254. doi:10.17730/0018-7259-74.3.243

Duran, A., Hoogendam, P., & Salazar, F. (1998). ¿Conflicto o concertación? La problemática del agua subterránea en el Valle Central de Cochabamba [Conflict or concertation? Groundwater problems in the Central Cochabamba Valley]. *PROCAMPO, 82*, 12–18.

FOBOMADE. (1999). *Conflictos ambientales en Bolivia; miradas, voces y sonidos* [Environmental conflicts in Bolivia; voices and sounds]. La Paz: Author.

García Linera, A. (2010). Estado y comunidad. Comentarios a la conferencia de Bernard Barraqué [State and community. Comments to the conference by Bernard Barraqué]. In F. Poupeau & C. González (Eds.), *Modelos de gestión del agua en los Andes, Actes y Mémoires de l'Institut Francais d'Études Andines*(pp. 33–39). Lima: Instituto Francés de Estudios Andinos.

García Orellana, A., García Yapur, F., & Quitón Herbas, L. (2003). *La « Guerra del Agua » Abril de 2000, la crisis de la política en Bolivia* [The Water War, April 2000, political crisis in Bolivia]. La Paz: FUNDACIÓN PIEB.

Gerbrandij, G., & Hoogendam, P. (1998). *Aguas y acequias; los derechos al agua y la gestión campesina de riego en los Andes bolivianos* [Water and canal; water rights and farmer-managed irrigation in the Bolivian Andes]. La Paz: Plural editores.

Harris, L., & Roa-García, M. C. (2013). Recent waves of water governance: Constitutional reform and resistance to neoliberalization in Latin America (1990–2012). *Geoforum*, *50*, 20–30. doi:10.1016/j.geoforum.2013.07.009

Hines, S. (2015). *Dividing the waters: How power, property and protest transformed the waterscape of Cochabamba, Bolivia, 1879-2000*. Berkeley: University of California.

Hommes, L., & Boelens, R. (2017). Urbanizing rural waters: Rural-urban water transfers and the reconfiguration of hydrosocial territories in Lima. *Political Geography*, *57*, 71–80. doi:10.1016/j.polgeo.2016.12.002

Hommes, L., Boelens, R., & Maat, H. (2016). Contested hydro-social territories and disputed water governance: Struggles and competing claims over the Ilisu Dam development in southeastern Turkey. *Geoforum*, *71*, 9–20. doi:10.1016/j.geoforum.2016.02.015

Hommes, L., Boelens, R., Harris, L., & Veldwisch, G. J. (2019). Rural–urban water struggles: Urbanizing hydrosocial territories and evolving connections, discourses and identities. *Water International*, *44*(2), 81–94. doi:10.1080/02508060.2019.1583311

Hoogendam, P. (2017). *Diseño conceptual del Componente riego de Proyecto Misicuni* [Conceptual design of the irrigation component of the misicuni project]. Cochabamba: Dirección de Gestión de Agua del Gobierno Departamental.

Hoogesteger, J., Boelens, R., & Baud, M. (2016). Territorial pluralism: Water users' multi-scalar struggles against state ordering in Ecuador's highlands. *Water International*, *41*(1), 91–106. doi:10.1080/02508060.2016.1130910

Instituto Nacional de Estadística. (2017). *Población de la Región Metropolitana Kanata* [Population of the Kanata Metropolitan Region]. La Paz: INE.

Kruse, T. (2005). La "Guerra del Agua" en Cochabamba, Bolivia: terrenos complejos, convergencias nuevas [The "Water War" in Cochabamba, Bolivia: Complex issues, new convergences]. In E. Garza Toledo (Ed.), *Sindicatos y nuevos movimientos sociales en América Latina* (pp. 121-161). Buenos Aires: CLACSO.

Laurie, N., Andolina, R., & Radcliffe, S. (2002). The Excluded "Indigenous"? The Implications of Multi-Ethnic Policies for Water Reform in Bolivia. In R. Sieder (Ed.), *Multiculturalism in Latin America*. (pp. 252-276). London: Palgrave.

Laurie, N., & Marvin, S. (1999). Globalisation, neo-liberalism and negotiated development in the andes: Water privatisation and regional identity in Cochabamba. *Environment and Planning*, *31*, 1401–1415. doi:10.1068/a311401

Linton, J. (2014). Modern water and its discontents: A history of hydrosocial renewal. *Water*, *2014*(1), 111–120. doi:10.1002/wat2.1009

Marston, A. J. (2014). The scale of informality: Community-run water systems in peri-urban Cochabamba, Bolivia. *Water Alternatives*, *7*(1), 72–88.

Ministerio de Medio Ambiente y Agua. (2015). *Diagnóstico del Sector de Riego con Enfoque de Cuencas* [Diagnostic of the Irrigation sector with focus on watersheds]. La Paz: MMAyA.

Molle, F., Mollinga, P. P., & Meinzen - Dick, R. (2008). Water, politics and development: Introducing Water Alternatives. *Water Alternatives*, *1*(1), 1–6.

Perreault, T. (2005). State restructuring and the scale politics of rural water governance in Bolivia. *Environment and Planning*, *37*(2), 263–284. doi:10.1068/a36188

Perreault, T. (2008). Custom and contradiction: Rural water governance and the politics of usos y costumbres in Bolivia's irrigators' movement. *Annals of the Association of American Geographers*, *98*(4), 834–854. doi:10.1080/00045600802013502

Perreault, T. (2014). What kind of governance for what kind of equity? Towards a theorization of justice in water governance. *Water International*, *39*(2), 233–245. doi:10.1080/02508060.2014.886843

Salazar, F. (2001). *Disputa y conflicto por el acceso al agua en el valle central de Cochabamba* [Dispute and conflicto over access to water in Cochabamba Central Valley]. Retrieved from https://www.nadir.org/nadir/initiativ/agp/free/imf/bolivia/el-agua.htm

Saldías, C., Boelens, R., Wegerich, K., & Speelman, S. (2012). Losing the watershed focus: A look at complex community-managed irrigation systems in Bolivia. *Water International, 37*(7), 744–759. doi:10.1080/02508060.2012.733675

Schultz, J. (2008). La guerra del agua en Cochabamba y sus secuelas [The Cochabamba water war and its consequences]. In Centro para la Democracia, *Desafiando la globalización: Historias de la experiencia boliviana.* (pp. 17-52). Washington: Centro para la Democracia.

Seemann, M. (2016). Inclusive recognition politics and the struggle over hydrosocial territories in two Bolivian highland communities. *Water International, 41*(1), 157–172. doi:10.1080/02508060.2016.1108384

SEMAPA. (2017). *Informe de Gestión 2016-2017; Rendición pública de cuentas* [Year report; public accountability]. Cochabamba: Author.

Spronk, S. (2007). Roots of resistance to urban water privatisation in Bolivia: The 'New Working Class' the crisis of neoliberalism and public services. In S. Spronk & J. R. Webber (Eds.), *Crisis and contradiction: Marxists perspectives on Latin America in the global political economy* (pp. 29–51). Leiden/Boston: Brill.

Swyngedouw, E. (2009). The political economy and political ecology of the hydrosocial cycle. *Journal of Contemporary Water Research & Education, 142*, 56–60. doi:10.1111/j.1936-704X.2009.00054.x

Swyngedouw, E. A. (1995). The contradictions of urban water provision; a study of Guayaquil. *Twpr, 17*(4), 387–405. doi:10.3828/twpr.17.4.c828066008888322

van der Zaag, P., Bolding, A., & Manzungu, E. (2001). Water-networks and the actor: The case of the Save River Catchment, Zimbabwe. In P. Hebinck & G. Verschoor (Eds.), *Resonances and dissonances in development: Actors, networks and cultural repertoires.* (pp. 257-279). Assen: Van Gorcum.

Zwarteveen, M. Z., & Boelens, R. (2014). Defining, researching and struggling for water justice: Some conceptual building blocks for research and action. *Water International, 39*(2), 143–158. doi:10.1080/02508060.2014.891168

OPEN ACCESS

Colonizing rural waters: the politics of hydro-territorial transformation in the Guadalhorce Valley, Málaga, Spain

Bibiana Duarte-Abadía and Rutgerd Boelens

ABSTRACT
This paper explores how, historically, the utopian thinking built into Spain's water policies has legitimized profound transformations of the Guadalhorce Valley's hydro-social territory (in Málaga), also justifying water transfers from rural to urban areas. It analyzes how the 'regenerationist hydraulic utopia' has been materialized through different 'governmentality strategies'. This intensified during Francisco Franco's dictatorship, decaying gradually into dystopias that, to this day, express profound socio-environmental impacts: dispossession, displacement, uprooting and breaking up local water governance institutions and practices. Meanwhile, the urban and tourism industries in Málaga have been strengthened by giving them priority for water supply.

Introduction

Málaga, located on Andalucía's sunny coast, is one of Spain's most visited cities because of its beautiful beaches, golf courses, water parks and excellent weather most of the year. Many tourists, mostly from Northern Europe, find this a very attractive place to live temporarily or for a vacation. In summer, Málaga's population soars; as does its demand for domestic water. Most water is taken from rural area uses in the Guadalhorce River valley, labelled as 'for human consumption', which is a priority use pursuant to Spanish legislation. However, much of this water also supplies profitable and recreational uses, such as watering gardens, showering on the beach, washing cars, private industrial uses and, partly, the boom in golfers' housing developments (cf. Delgado & Del Moral, 2016; Villar Lama, 2013). As we document, water demands for urban uses and tourism are met at the expense of rural inhabitants of the Guadalhorce watershed, who have to ration irrigation water during droughts and, in emergencies, lose their crops and fruit trees.

To understand this situation, beyond rhetoric about 'national interest' or 'top priority for human consumption', this paper analyzes Spain's hydro-territorial policies historically, with specific attention to rural–urban water transfers from the Guadalhorce Valley to Málaga over the last 100 years (cf. the introductory paper to this issue, Hommes, Boelens, Harris, & Veldwisch, 2019). Therefore, we first describe 'the regenerationist dream and hydraulic utopia', intellectually led by Joaquín Costa, beginning late in the 19th century. It proposed to revive the nation by giving top priority to expanding hydraulic works and governance through watersheds (e.g., Fernández Clemente, 2000; Mendoza, 1992; Swyngedouw, 2007, 2015). Second, we examine how this regenerationist water-based utopia is materialized in the Guadalhorce Valley through the different governmental strategies implemented in the 20th century. We analyze the famous engineer Rafael Benjumea's hydraulic works, later intensified under the Francisco Franco dictatorship (1939–75) (see also Boelens & Post Uiterweer, 2013). The *Política Hidráulica* re-ordered the Guadalhorce territory and its inhabitants' lives to undertake projects on behalf of the national interest: urban electrification and water needs for industrial and domestic uses.

We analyze how Joaquín Costa's hydraulic dream has gradually become a dystopia for many. This paper compiles historical information, technical reports, hydrology plans, biographical narratives, articles and scholarly studies. We articulated this information with field visits and interviews, conducting 49 semi-structured interviews and 12 collective meetings with key stakeholders (14 interviews with farmers, 10 with official agents, five with water inspectors, five with residential tourists, five with displaced families, four with design technicians and six with elderly inhabitants). Next, we recorded several in-depth life stories of the Guadalhorce Valley's inhabitants. The research included in total seven field visits throughout the period 2015–17. We triangulated our data using the above-mentioned diversity of research methods and interviewing different categories of actors. Further, the researchers have guided 12 student groups collecting data and validating and updating the growing data set. Finally, the research data and conclusions were discussed and checked with officials and the Guadalhorce Valley water users.

The next section discusses the conceptual approach toward analyzing the problems of rural-to-urban water transfers. Applying a political ecology approach, our analysis focuses on understanding the Guadalhorce Valley as a hydro-social territory (Boelens, Hoogesteger, Swyngedouw, Vos, & Wester, 2016). The third section documents how integrated management of Guadalhorce's land and water harks back to Arab times – with *huerta* homesteads and croplands, customary irrigation systems and self-governed communities. This socio-technical, idealized background constitutes the historical fundament for configuring 20th-century utopian–modernist imaginaries and dreams as well as Fascist hydro-territorial planning policies. We explain hydraulic utopia's rootedness in the national political context, and show the valley's hydro-territorial transformations throughout history: its foundation being the damming and diversion of the river to modernize and bring progress. The fourth and fifth sections illustrate the otherness of hydraulic utopia: the dystopias that many rural people have experienced, and sometimes still have to face. Here, Spanish scalar politics are steamrollered by 'national well-being and public interest' (for comparison, see, for instance, Harris, 2002; Hommes & Boelens, 2018; Kaika, 2006; and Swyngedouw & Williams, 2016). These latter imaginaries and discourses legitimize rural water transfers not only to domestic users but also to capital generating sectors: the tourism industry and urban elites. Diverse 'governmentality strategies (Foucault, 1978/1991) will be analyzed to understand how these water transfers from rural zones are normalized, accepted and legitimized. Nevertheless, recent hydro-modernist transformations in the Guadalhorce Valley are also contested, in particular by citizens' alliances and alternative territorial modes of valuation.

Hydro-territorial transformation and governmentality: a utopian/dystopian conceptualization

This paper examines how utopian hydraulic policy 'for all' (rural and urban uses) ends up prioritizing powerful urban interests, turning it into a water control dystopia for some rural livelihoods. Since Plato constructed the ideal State in *The Republic* (380 BC) and above all Thomas More's *Utopia* (1516), there has been a longstanding philosophical and literary tradition resulting in hundreds of manuscripts with policies and designs that seek to materialize 'the art of utopian governance'. These utopias characteristically claim to rescue society from its structural chaos and deep-rooted crisis. They project ideals evoking imagined worlds for a better future (Achterhuis, Boelens, & Zwarteveen, 2010). A major historical case is the sociopolitical, intellectual movement of 'Regenerationism' in late 19th-century Spain. After losing its last colonies in the Philippines, Puerto Rico and Cuba in 1898, Spain ceased to be a global empire. This profound crisis – culturally and symbolically imagined, economically perceived and politically constructed – resulted in the quest for a new national identity and political/economic modernization. The regenerationists' movement promoted the ideological foundations for Spain's new cultural, political and economic direction (Maurice & Serrano, 1977; Mendoza, 1992; Ortí, 1984; Swyngedouw, 2007). Formed by idealistic intellectuals, writers, politicians and also deeply influencing technocratic and culturalist modernizers, this broad progressive, modernist movement used apocalyptic predictions to sustain its desire to radically reshape society for national improvement, importantly based on, among others, the following axes: hydraulic mastery; an increase in agricultural production; attention to local knowledge systems, customary laws and practices; and decentralized management (Boelens & Post Uiterweer, 2013).

Achterhuis (1998) defines utopia as a 'feasible' society that can be changed and neatly perfected by its creators and founders. Therefore, 'utopia' does not relate to individual dreams and lives but to overall constructs of a 'new society'. Both literary works and historical political experience have shown that, worldwide, founding and creating a 'utopia' requires radically breaking with the past and with the existing order to undertake a new present – to build the desired society (Achterhuis, et al., 2010, p. 29). In practice, these breaks are framed in designing drastic reconfiguration, violent interventions, exclusion and repression of dissenting speech and thinking, and destroying the old society's structures and cultural norms (Gray, 2007; Kumar, 1987; Levitas, 1990). Usually, building the desired society, materializing utopian ideals, results in violent dystopias.

This paper addresses the 'hydraulic utopia' pursued by the regenerationist movement and its successors in Spain (Boelens & Post Uiterweer, 2013; Ortí, 1984; Swyngedouw, 2015). More specifically, our contribution is that we address the link between territorial planning and water governmentality endeavours: the socio-technical interactions aiming to control and transform the Guadalhorce Valley's dynamics to put in order 'hydro-social territory' and the inhabitants' ways of life – aiming to change humans and territory at once. By using utopian and governmentality theory, we show how human relations, customs, habits, ways of acting and thinking are manipulated to control water flows and adjust territories (Boelens, 2017; Duarte-Abadía, Boelens, & Du Pré, 2019; Lopez-Gunn, 2009).

Governmentality means understanding the ways in which technologies, truths, discourses and rationalities act and are made to act, consciously or not, as forms of power to conduct people's thoughts and practices. We shall explore three ways in which 'the art of governance', or 'governmentality' ('government rationality'; Foucault 1978/1991, 2008), is expressed by developing utopian water policies. The first is sovereign, grounded in safeguarding common well-being and public utility by means of installing and claiming respect for laws and formal state policies. It is imposed by the sovereign entity that threatens with and applies the (self-) legitimized use of violence. The second is regulation through discipline, which normalizes thoughts, behaviours and practices through subtler, less visible moral mechanisms, assembled into a heterogeneous array of discourses, institutions, laws, technological designs, administrative measures, scientific truths and moral values (Foucault, 1978/1991). The third refers to neoliberal governmentality, creating structures for market forces to organize societal relationships; it enables market rationality to operate in different walks of society, generating free circulation of human and non-human commodities (Vos & Boelens, 2018). This motivates and manipulates individuals, viewed as individualistic, rational stakeholders, by creating economic incentives (Fletcher, 2010).

These strategies to drive behaviour ('the conduct of conduct'; (Foucault, 1978/1991) materialize utopian thinking and imaginaries socially and physically. Assembling financial resources, institutional practices, hydraulic infrastructure, legal frameworks and human behaviours shapes spatial configurations of territory and control over water (Baviskar, 2007; Meehan, 2013; Swyngedouw, 2015). Thereby, we use the notion of hydro-social territories, referring to the contested imaginary and socio-environmental materialization of spatially bound multi-scalar, socio-natural networks. Thereby, governmentality projects aim to mobilize and align hydraulic infrastructure, water flows, cultural institutions, political practices and economic relations to create 'convenient', dominant order: to control nature and societies, at once, through water (for conceptualization, see, for example, Boelens et al., 2016; Hommes & Boelens, 2017, 2018; Linton & Budds, 2014; Marks, 2019; Seemann, 2016; and Swyngedouw & Boelens, 2018). Hydro-social territories are continually disputed: hegemonic power shapes them, as do contestations by population groups that are excluded, marginalized or affected by the materialization of utopian projects.

History of hydro-territorial transformation of the Guadalhorce basin: modernizing water and people

In this section we outline regenerationists' hydro-political thoughts and plans to revive the country after Spain lost its last colonies and entered a deep crisis. They aimed to abolish oligarchic relationships and the deeply cumbersome living conditions of the peasantry, among others and in particular, through reorganizing water control. This national 'utopian hydraulism' landed in the Guadalhorce Valley through the hydropower dam and governance institutions introduced by politician and engineer Rafael Benjumea, later public works minister. He deeply transformed the Guadalhorce Valley's hydro-social territory and influenced the future of Spain's socio-natural/technopolitical development.

Joaquín Costa's hydraulic utopia: equitably sharing water, redistributing land and social justice

In the late 19th and early 20th centuries, continued land decommunalization and peasant marginalization intensified rural impoverishment constituted the background of the thought and works of the regenerationist movement, led by Joaquín Costa. Regenerationism reflected his intense desire to restructure Spanish society, especially empowering small farmers by territorial reorganization, hydraulic modernization and water redistribution, to encourage agrarian productivity (Gómez Mendoza, 1992; Ortí, 1984).

Costa's hydraulic utopia viewed Spain as a shattered country of poverty and hunger, producing too little food and being politically stagnated because its power structures – oligarchy and elitism – prevented any transformation (Costa, 1911, 1967). Deeply rooted in modernization ideologies, this crisis called for 'regeneration' based on simultaneously 'improving the land and the people of Spain', resulting in a 'new man'. This entailed a profound change in inhabitants' education and mindset.

Costa's educational proposal to install an irrigation doctrine and hydraulic reterritorialization relied on positivistic science. He considered hydrographic and orographic data to control water to be fundamental; along with folk wisdom whenever rationalized, unified and formalized as shared heritage (Maurice & Serrano, 1977). Along with new flows of water, such knowledge had to spread nationwide. The key regenerationist idea was rather than colonizing distant lands, to begin *colonizing their own country*. In this imaginary of 'inward colonization' to 'revive' the country, building water projects would play a fundamental role: to transform nature, regenerate soil fertility, foster human intellectual production and overcome economic difficulties (Boelens & Post Uiterweer, 2013; Swyngedouw, 1999, 2015).

This in-country colonization viewed the state as a representative of the public interest, an ideal entity for constructing a perfect harmony between the common interest and those of irrigators and the rural population. The state would be above the interests of political parties and social classes. Regenerationism promoted *national hydraulic solidarity*, aware that large landowners would have to make an effort to work for this 'common well-being' (Fernández Clemente, 2000). To end the feudal system, Costa proposed reforming the oligarchy and elite powers, moving toward the rule of law – not on the basis of a class struggle and Socialist revolution, but through consensus-based reform. The reformist intention was not to take away from the few to give to the many, but simply to give to [the] many, by distributing public lands and extending irrigation to all irrigable drylands. This would, in practice, obviously benefit large landowners (Costa, 1967). Therefore, water should be nationalized and state controlled, defending the common well-being and contributing to the nation's overall development (Ortí, 1984; Swyngedouw, 2015). Building canals, irrigation systems and reservoirs under Costa's hydraulic policy would enliven the country's economy from a social and technical standpoint.

To enact hydraulic policy, ironically, Costa and other regenerationists advocated a government of 'action men' to break the political stagnation caused by regional elite rule and the undemocratic monarchies of Alfonso XII and XIII. This claim for a shift in leadership sought to enhance participatory structures and agrarian collectivization – regenerationist ideology was strongly connected to the Republican democratic objectives, such as 'the free institution of teaching', committed to promote and strengthen the capacities of the peasant communities against the power of rural caciques (chieftainship) and oligarchy. Maurice and Serrano (1977) analyze this regenerationist demand for a strong chief of state to be both authoritarian and populist: authorized by the community at large, familiar with the people's anatomy, embodying common interests, representing the national identity and forging harmonious covenants with the people. This way, Costa introduced the need for a 'surgical policy, which must be wielded personally by an iron-fisted surgeon' (Costa, 1967, p. 86). However, he was also aware of the need to limit not just oligarchic but also state powers, for which he proposed forming political–administrative organizations to govern river basins, involving users in decision-making.

Controlling water to regenerate territory and humans: the count of Guadalhorce

The thinking of Costa and regenerationism greatly influenced 20th-century politicians and professionals. This included Rafael Benjumea, who directed the mega-dam construction on the Turón River in Guadalhorce (1914–21), and was Minister of Development (1926) and of Public Works (1937). To understand current water-territorial planning and water governance, this section details historically how the hydraulic utopia began transforming hydro-territorial relationships in the Guadalhorce Valley.

Crossing through Málaga province, the Guadalhorce River ('Wheat River' in Arabic) is 154 km long (Agencia Estatal, 1961). Before emptying into the Mediterranean Sea, on the 'Desfiladero de los Gaitanes', two major tributaries join, the Turón and Guadalteba rivers. From the eighth to the 15th centuries, Arabic culture left its legacy in irrigation canals. Some 4000 hectares were farmed by small landowners, rural communities and large landholders, until the mid-20th century. Boelens and Post Uiterweer (2013) document that farming communities in Guadalhorce managed their socio-hydraulic systems collectively; by rebuilding small rustic dams (*azudes*) every year, they maintained their small-farm irrigation. Communities were self-governing, with no official registration, and independent of public administration; they divided up roles and tasks to manage their irrigation systems and meet irrigators' needs. Collective water management required cooperative work, which generated values of cohesion and solidarity among Guadalhorce farmers.

However, contemporary engineers felt Guadalhorce was full of water uncertainties, too meagre or too torrential flows, even jeopardizing Málaga (Brotons & García, 1999; Martín-Gaite, 2003). Following the general national plan of irrigation canals and dams of 1902 (Cantero, 1995), a catalogue of 296 waterworks, Rafael Benjumea set about fixing these problems and stimulating internal development of Málaga society by waterworks (Fernández Clemente, 2000). Between 1903 and 1905, starting the Guadalhorce Valley's radical techno-political transformation – and emblematic for the country's new, modernist hydro-policy century – Benjumea built a large hydropower plant at El Chorro, supplying electricity to Málaga.

However, the Guadalhorce River's summer flow could not cover the expanding electrical demand, and in winter this facility did not protect the city from potential flooding, so Benjumea proposed the ambitious project of damming the entire Turón River (Figure 1). Nevertheless, this mega-project would be justified solely by intensifying agricultural production. When, in 1911, the waterworks development law was enacted – helping irrigator communities by paying half the costs of improving infrastructure and expanding irrigation – this made it possible to begin the large project.

For this reason, despite his main interest in meeting the citizenry's energy demand, Benjumea promoted a sensitization campaign to convince rural inhabitants of the project's benefits. However, in the Guadalhorce Valley, most farmers located in the upstream area were too small to co-finance and not interested because they already had their traditional, well-functioning systems. Nevertheless, Benjumea agreed with the large landowners, located downstream, to form an Agrarian Union to tap into the government funding and build the large reservoir (Mártin-Gaite, 2003, pp. 51, 52).

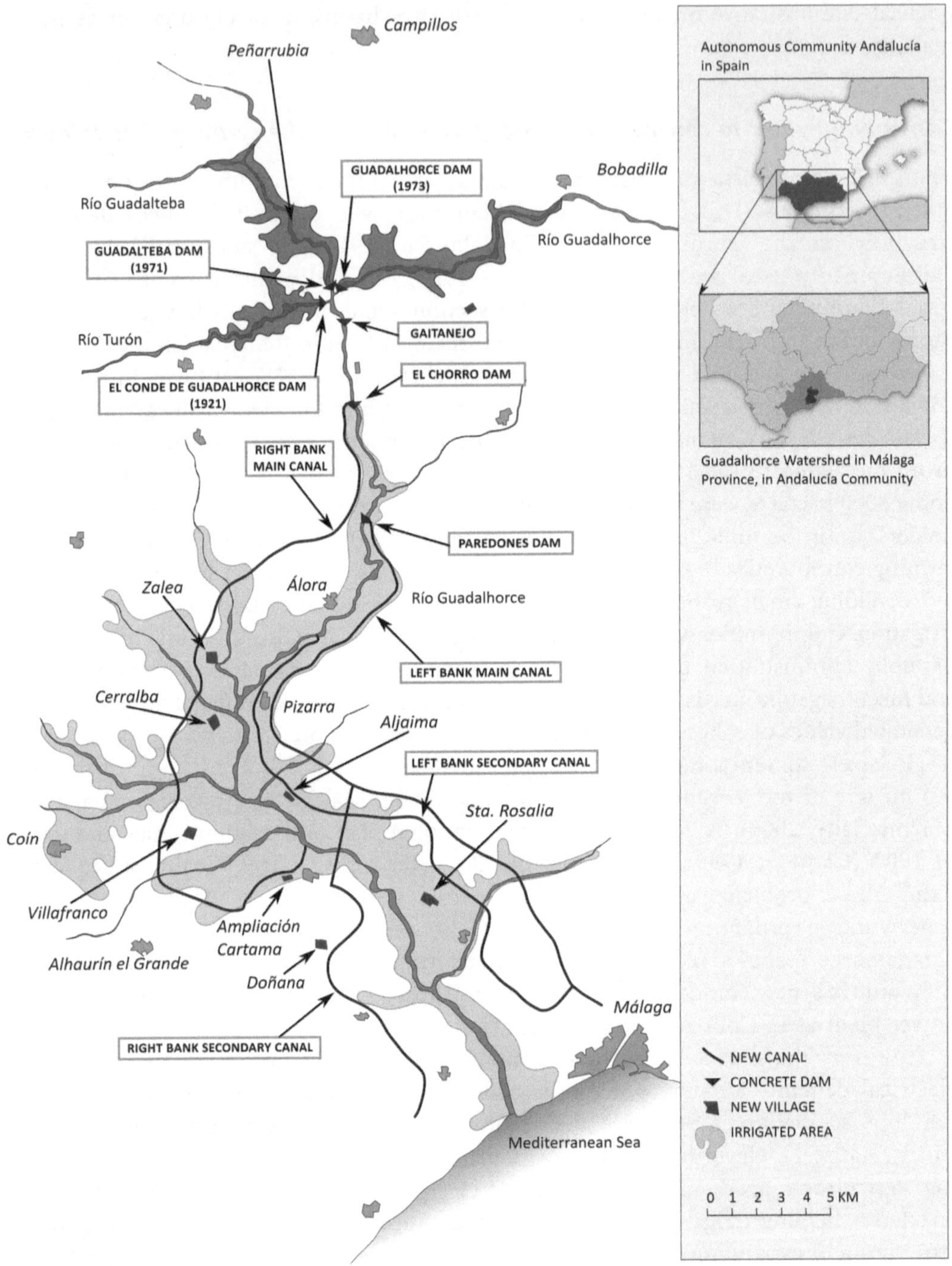

Figure 1. Guadalhorce Valley: reservoirs, irrigation system and rural towns.
Source: Adapted from Boelens and Post Uiterweer (2013).

Built from 1914 to 1921, its main purpose was to supply electricity, buffer flooding and irrigate 13,000 hectares in the valley for towns in the Málaga basin. The majestic dam, 72.5 m high and 86 hm^3 (86 million cubic metres) in capacity, was celebrated and inaugurated by Alfonso XIII, who made Benjumea the 'Count of Guadalhorce'. This later became the name of the dam on the Turón River. In the 1960s, this damming project enabled the 'Coordinated Irrigation Plan of Guadalhorce'.

The reservoir stopped the river flow, but was unneeded for most small farmers' purposes because even in the dry summers they could water their crops by catching water by small dams on the Guadalhorce River that they made themselves (Boelens & Post Uiterweer, 2013). Even so, the reservoir was legitimized by playing a 'general interest' social function. Therefore, engineers' works performed a technical, social and patriotic mission (e.g., Brotons & García, 1999) that would forge Spain's way out of its crises, shedding a light in the darkness.

To continue regenerationist hydraulic policy, in 1926 Benjumea was appointed Minister of Development by dictator Primo de Rivera, and created the national water policy administration through the internationally acclaimed hydrographic union confederations (Confederaciones Sindicales Hidrográficas). These agencies understood river basins as the basic unit of water development and management, proclaiming a utopian-inspired (participatory, democratic, autonomous, decentralized) approach (Mülberger, Vilaró, Tirado, & Domènech, 2007). These were in charge of connecting and integrating all water infrastructure, and territorial transformation, contemplated in the general plan of 1902. Hydrographic union confederations would materialize the regenerationist utopia through a participatory bottom-up structure aimed at abolishing the power relations between the feudal, rural oligarchy that obstructed the implementation of an agrarian reform and land distribution (Frutos Mejias, 1995; Sanchis-Ibor, 2012).

Hydraulic dystopia

Years later, for many rural inhabitants of the Guadalhorce Valley, however, building this hydraulic utopia became a hydro-social tragedy. Paradoxically, the regenerationist hydraulic policy dream was adopted by Franco's dictatorship and had contradictory effects to those initially proposed by Costa. This section will portray the diverse governmental strategies that have transformed the Guadalhorce hydro-social territory, leading to displacement, dispossession and relocation.

Materializing hydraulic utopia: Generalísimo Franco, hydraulic surgery and dystopian transformation of the valley

When the Phalangist nationalist front, led by Franco, won in 1939, this meant interconnected repression of water flows, freedom of thought and hydro-social identities of the Guadalhorce Valley communities. Seeking to align water and human currents under a single repressive system, the military government made the dreams and thinking of hydraulic regenerationism come true, be it with outcomes different from what they had hoped for.

The national hydraulic integration and unification was mobilized by power alliances among the military, Church, industrial bourgeoisie, large landholders and the government agency (cf. Swyngedouw, 2007). Illustratively, the Corps of Engineers came to occupy most political offices, directing different ministries. Large landholders supported the Franco regime, which promised not to change the property status quo and rather proposed to build new towns to put rural inhabitants into order and under discipline.

The media glorified the dictatorship, stating that the Fascist regime (1939–75) showed concrete results, implementing the long-awaited water policy – the 180 reservoirs in 1939 swelled to 800 by 1975 (Orti, 1984; Swyngedouw, 2007).

Meanwhile, the nation's hydro-territorial geography was redesigned through large-scale violence (Swyngedouw, 2015; Swyngedouw & Boelens, 2018); people with ideologies deviating from that of the regime were oppressed and many even enslaved – well known are those hydraulic works in other parts of Spain that were built by political prisoners, 'Franco's slaves' (Acosta Bono, Gutierrez Molina, Martinez Macias, & Del Rio Sanchez, 2004; Camprubí, 2013; Lafuente, 2002). Violent practices, as a whole, comprised the governmental strategies interconnecting the country's hydrological systems, also transforming the Guadalhorce Valley. For example, as a water inspector stated in an interview:

> During the civil war towns such as Peña Rubia, located in the upstream of the Guadalhorce river, were emptied, people fled from the Nationalist and Republican front. Many were killed and shot in Alora, there was a lot of displacement among different towns in Guadalhorce. (7 October 2016)

Feelings of repression, suffering and – till these days – acts of silence and the wish to forget, in order to carry on, is the behavioural pattern observed in the older residents of this valley.

Forceful hydro-territorial transformation and land dispossession became generalized practices under Franco. Different from Lorenzo Pardo's 1933 more participatory National Water Plan, in 1939, the General Public Works Plan stressed state intervention to pursue and integrate waterworks and merge hydraulic policy with agricultural transformation through colonizing dry land, *marisma* wetlands turning them into irrigated cropland (Fernández Clemente, 2000). Later, in 1949 – the bitter legacy of the regenerationist dream of domestic colonization – the laws on colonizing and distributing property in irrigable zones and on forced expropriation were enacted. These entitle the government to expropriate to build waterworks. Expropriating included forming 'colonization towns' (*pueblos de colonización*) and a programme of agricultural indoctrination through The National Colonization Institute (INC).

The 1939 law encouraged new waterworks and territorial plans in the Guadalhorce Valley to further discipline river flows and rural residents and, above all, guarantee the water supply for the growing city of Málaga. As Luis Morales, former director of reservoir building and management, explains:

> The dictatorship's technocratic governments began to see that this zone had great tourism potential, but required guaranteed water supply. On this basis, the Coordinated Irrigation Plan of Guadalhorce was created to regulate the Guadalhorce and Guadalteba Rivers. [...] The fundamental criteria were, first, to guarantee water supply for Málaga, with a legally allocated flow rate of 1500 l/s from the Guadalhorce River. But there was no infrastructure to handle it. This was the fundamental reason to build the Guadalhorce–Guadalteba dam. The second aim was to expand irrigation, followed by secondary goals such as electrical energy. At the same time, the colonization towns were built, seven or eight new towns. (9 October 2016)

The 'colonization towns' of Villafranco, Doñana, Cerralba, Zalea and Santa Rosalía (Figure 1) were to house the people displaced to build the new waterworks. Agricultural indoctrination programmes were brought to the Guadalhorce Valley to facilitate 'exchanges of knowledge': ideas that Costa had dreamed of and would lead to one of the nation's greatest agrarian transformations. The same occurred with the hydrographic union confederations, whereby their participatory spirit was completely removed by the Francoist system to become controlled by top-down civil engineers' technocracy, re-named as hydrographic confederations (*Confederaciones Hidrográficas*). In this way, large landowners and energy enterprises' interests were protected.

Works on Guadalhorce and Guadalteba reservoirs began in 1966 and ended in 1973, with a total capacity of 328 hm^3 to water new croplands. Nevertheless, the main purpose was to cover Málaga's drinking water supply. In the 1960s, most of the ditches were also made, but only 9000 of the planned 20,000 ha were irrigated (and from which 4000 ha were already covered by the traditional Arab systems). The design allocated 92.5 hm^3 annually to irrigation (the amount varying, as available) and 9.5 hm^3 (fixed and guaranteed) to supply Málaga (Calvo, 1973). However, Roberts (2002, p. 154) states that, after three decades, 50–60 hm^3 annually are used for agriculture and 47 hm^3 for domestic and commercial uses. When water is scarce, the water supply for Málaga has total priority.

The mega-reservoirs neatly fit with Franco regime's territorial reordering, based on 'development poles': Franco imagined that building reservoirs all over Spain would make it possible to generate enough hydropower to set up urban–industrial centres in 'backward regions'. Andalucía was to be the tourism development pole. The regime favoured urban–industrial development far more than agriculture (Del Moral & Saurí, 1999).

Governmentality and loss of collective control

Franco's policy gave prime urgency to the Guadalhorce and Guadalteba dams: engineers were given fewer than four months to submit construction plans. Luis Morales: 'In Franco's time, there was no opposition, so of course a minister's decision was irreversible' (9 October 2016). This allowed no suitable ecological or geological studies, resulting in works with geophysical faults that supply brackish water to this day, far saltier than allowed for drinking or even irrigation water. Further, the modern system loses too much water from its elevated, prefabricated concrete canals. They were made of inadequate materials; traditional ditches have better consistency. The former reservoir operations director says the scientific and technical errors committed in the Guadalhorce irrigation system resulted from the authoritarian regime commanding designers and builders: 'They were ruinous from early planning through the present day, technically, in totally inadequate infrastructure characteristics, and socially, with plots designed that were too small for a household's livelihood' (20 October 2016).

The less visible, dark side of the hydro-territorial transformation is obvious in what small-farm irrigators received – and also in what they have lost. The Franco regime's water transformation meant destroying the rustic dams on the Guadalhorce River and small farmers were forced to join the large hydrographic confederation. These confederations ultimately combined the Corps of Engineers and official entities for the sole management of the watershed, replacing and governmentalizing local self-governance systems and their traditional collective authorities (Boelens & Post Uiterweer, 2013).

During the transformation, the private, market-based land tenure structure was strengthened, while dissolving collective arrangements and community areas: by building the waterworks, imposing discourses about general public interest, and by implementing irrigation doctrine, colonization laws and in-country repopulation. Fernández Clemente

(2000) says that the INC commonly bought land from large owners to implement waterworks, paying high prices, and letting them keep most of their properties. The colonization towns also assured them of plenty of cheap labour, disciplined by the Franco regime (see also Closas, 2018). Water interventions on the Guadalhorce River since the early 20th century have brought the watershed and its flow patterns under technical/technocratic domination, restricting freedom of access and river-based social relations. Juan Francisco Martín, a water inspector, explains: 'The Guadalhorce River was chopped by the guillotine of the upstream dams and La Encantada' (22 June 2015).

In the upper basin, water is dammed in three large reservoirs, so the Guadalhorce's natural dynamics have vanished, and inhabitants' relational bonds with the river and each other have been trampled. These days, the banishment of these bonds is seen in irrigators' organizational difficulties in cooperating collectively to manage the irrigation system, after having been excluded from designing and planning it. In general, each user acts individually to obtain water, and many have lost their feelings of rootedness and belonging.

Gradually, the public assets of the Guadalhorce Valley have been fenced in: first, riparian and community areas were seized for 'public-interest uses', to build waterworks and expand irrigation; the agricultural failure was then rescued by the growth of capitalist residential tourism, generating further land subdivision, sales to foreigners and individualization of land and water. In the Guadalhorce Valley, English and German gated colonies have formed, interacting minimally with local inhabitants and knowing nothing about the water distribution operating rules. Many of them do not respect irrigation scheduling and, when they see water running through nearby ditches, they take it to fill their swimming pools or water their gardens. Morales: 'The water from the reservoirs, once regulated for irrigating trees, still waters some trees but fundamentally fills swimming pools. In fact, consumption peaks on Sundays, when people change out the water in their pools' (5 October 2016).

Building dams, displacement and rootlessness: efforts to continue living

The hydraulic dream works that were materialized under Benjumea, and then intensified under the Franco regime, adversely undermined the social justice aims of regenerationism's water utopia. In the Guadalhorce Valley, reservoirs have flooded rural communities, displacing families elsewhere. Colonization and forced expropriation laws, plus the agricultural indoctrination, removed people from their territories and destroyed their roots. As water inspector Martín puts it: 'The colonization towns are parallel to dam construction. The idea of the Guadalhorce irrigation system was to "fix" the population […]' (22 June 2015).

His colleague, Manolo Rengel, confirms this: 'Territorial planning under Franco was to colonize, locating people strategically in the territory […] whenever someone stood up against Franco ideology in one territory, they were neutralized and taken somewhere else' (22 June 2016). This way, Franco's strategy delocalized, governmentalized and disciplined territory, uprooting and exterminating culture, to mould a new society according to the dictates of the Fascist hydro-territorial regime. Cristina López, who lives in the valley, explains:

> They brought many people from Valencia to the colonization settlements, supposedly to teach, but mainly I think to uproot people from their land. And since no protest was allowed, this weakened people ideologically and morally. Because normally, if you are not born in that particular place, if you have no roots there, no ancestors or relatives, you won't defend that place; you don't feel the same interest. (22 June 2016)

Literature, furthermore, agrees that Francoist inner colonization tried to create settlements loyal to the regime. These were strongly ideologized (Closas, 2018). Often, selected (very poor) people among the applicants (mainly peasants without lands or people forced to move because dam projects flooded their homes) became the members of these settlements. A 'good behaviour' certificate signed by a priest or a clearance certificate were regularly required (Lafuente, 2002).

Two life stories illustrate how the new hydraulic infrastructure changed the lives of Guadalhorce Valley rural families in a dramatic way.

The case of El Chorro: examining displacement from the inside

The Guadalhorce River cuts through the Penibetic peninsula, forming a narrow canyon called the Garganta del Chorro or Desfiladero de los Gaitanes. Past the canyon is the Tajo de la Encantada (or El Chorro) reservoir, with its main canal intake to supply water both to Málaga and to water the valley's crops (Figure 1). The reservoir and its pumping system is part of the Sevilla Electric Company power plant. This plant adds no generation capacity, but reflects capitalist competition between electric companies to produce peak-hour energy for the city rather than any actual need (Lara, 2002). The facility replaced the former El Chorro power plant and intake and flooded the local community. Utopian planning favoured urban downstream interest but created dystopia for the rural communities.

Andalusian families had lived there, growing dryland crops and irrigating their *huerta* multi-cropping gardens, and raising livestock. One was the Pérez and Rengel family. In the 1960s, brothers Antonio and Manuel Rengel married sisters Pepa and Teresa Pérez. The two brothers worked together, Antonio handling field activities and Manuel selling and distributing their produce. Manolo Rengel, the younger son of Teresa and Manuel, was born in Cortijo del Chorro in 1966, and remembers playing with his cousin, Miguel (son of Pepa and Antonio), while the ditch to catch water from the river was maintained. 'My father and uncle, with two or three other men, were cleaning the ditch [...] the azud dam was repaired, and they began from there every year, maintaining the first stretch of the ditch' (Manolo Rengel, 6 October 2016).

Different rumours warned that their land would be swamped one day, but were disregarded as unconceivable. Then, in 1972, employees of the Sevilla Electric Company turned up, offering to buy their homes. Cristina López, Manolo's wife, tells about that injustice: 'they expropriated for the "fair price", which was the price that the Government considered fair [...]' (22 June 2016). Aunt Pepa continues: 'we had no chance to negotiate the price – the Government came along, and said "this is worth this price, and you are leaving", regardless of what we said' (6 October 2016). Otherwise, they would receive no payment, and the machinery would demolish their homesteads anyway. Because of the dictatorship, there was no way to negotiate, much less protest. Aunt Pepa: 'There were never any complaints afterwards, because everything was settled for the Nation's best interests, for the hydropower industry' (6 October 2016).

Manolo cannot get the memory out of his mind: 'In 1974 the expropriation began. I still remember when they came in with machinery to tear up the groves we had tended so lovingly. [...] The expropriation, dam-building, uprooting people from their land and customs, was all traumatic' (22 June 2016). This episode marked his childhood and his life. Manolo tells how they lost and longed for their land, when he lost his father, who was already ailing, but lost the strength to carry on.

The shared memories heavily impacted every family there, as Pepa states: 'I remember seeing the bulldozers there, waiting to clear out our trees, laden with lemons – it was so sad, and such sorrow. While the machinery razed everything, all the children cried like babies; it was terrible' (6 October 2016). Juanita, their neighbour at the time, says that: '[...] The Agroman construction company destroyed everything to build the project. They made the upstream dam for the power plant, too. They even destroyed all the archaeological Arab churches and homes, and the ancient Bobastro fortress' (5 October 2016).

Manolo's mother was 35 years old when uprooted from her country way of life. She has never gotten used to the new places where she has had to live, between Málaga and Alora. Teresa feels the dam was a way to divide and remove people who wanted to remain in El Chorro, especially when she was widowed with five children after the expropriation.

> Coming to the city, where I knew no one, going into a house without any idea where there might be a school, or where anything was, forced me to wander the streets, looking [...] it was so hard to adapt to this new life, where everything was unknown – everything. (6 October 2016)

As Manolo put it, 'we were displaced in time and in space' (6 October 2016). Antonio, Manolo's uncle, after the expropriation, began suffering from nervous disorders because of the disorientation caused by the new place. Pepa and Antonio lived only five years in Málaga and in 1977 returned to the countryside, buying land in the Guadalhorce Valley, near Alora. Now Antonio is 84 years old and continues caring for the land they purchased. Manolo tells that 'like my father, anywhere my uncle was, he was displaced, disconnected' (8 October 2016). Even after so many years, he never recovered what he had worked for in the Chorro Alto fields: 'from a productive farmer, he was reduced to subsistence farming' (22 June 2016).

Teresa, who was paid the 'fair price' compensation, says the amount she was forced to accept could never replace what was taken away, which she has longed for all her life. When the El Chorro reservoir's waters recede, they uncover the past they flooded. Pepa feels grief when she sees this: 'sometimes it is very empty, and we can see the walls of our home [...] we see the bridge crossing the river, the ruins of the factory, the other homes here [...] they show what we had' (14 January 2017). Building the dam, to modernize the region, destroyed their future: 'it was unnecessary for us. We had everything: our groves, the fields where we planted and lived' (14 January 2017).

The Rengel and Pérez family still visits El Chorro. The home of Manolo's maternal grandparents was not flooded, so they gather there for family occasions. The displacement trauma affected their parents, but memories of this dispossession also have been passed down to the following generations. They identify profoundly with a town that no longer exists. Manolo feels that his land was flooded because his family's rights were seized and turned over to outside interests, to faceless modernization. 'We have never been able to get back to what we had before. We wanted to recover it, to re-establish what we remembered, but it was impossible. That was all dramatically changed forever' (22 June 2016).

Now, Manolo is a water inspector for the Guadalhorce irrigation system. He distributes water among users who irrigate and townspeople in Málaga, from the intake at the Tajo de la Encantada reservoir, at the dam that flooded his family's dreams and part of their story. Day after day, he works to bring justice to small farmers in the valley, despite official priorities of 'water for Málaga'. His work helps provide the means for the livelihoods of farmers and labourers still living in the countryside. He is also a local player in the fight to 'bring the river back to life' by providing the minimum ecological flow that will keep the river and its landscape alive.

The case of Peñarrubia: the voices of those who were silenced

> The State should have been more generous with these people whose lives it destroyed without even asking. Perhaps homes, land, and so on were paid for, but the moral damage, being able to say 'I was born in this town and grew up here, and I like it – why should I have to move somewhere else?', I think that value was never compensated for. (Luis Morales, former reservoir manager, 5 October 2016)

Peñarrubia was founded in 1500. In 1970, its population counted 1750 inhabitants; its area was 3845 ha, mostly for rain-fed agriculture, the people's livelihood (Figure 1). Most land was divided into small plots. Peñarrubia farmers had their own gardens for self-supply and rounded out their income by working as farmhands for large landowners.

The Guadalhorce River was fundamental for families' lives. They brought water to irrigate pastures along the river, watered their animals, washed their laundry, fished and everyone bathed in the river. In Peñarrubia they spent most days working in their fields and their leisure time also focused on their town. Juan Mora, displaced from there, tells us:

> On Rosary Street, the courtyard with the same name had the grocery market. Every woman in town would shop there. At the market, Antonio sold meat, Paca sold fish, Rosarito around the corner sold vegetables and Rafaelito did, too, and Aníbal Mendosa sold the freshest fruit. […] I loved our life in Peñarrubia. When we would get back from the fields, we would wash up, change clothes, and go up to the plaza. There was the church, city hall, Pepe's Bar, Juanito Corral's, Mendoza's, the movie theater, Cristóbal Pozo's bakery, the kiosk, and a place at the end of town where we loved to stroll. […] It was like a fair; every woman in town would go shop there in the morning. A precious setting, which we have lost. (11 October 2016)

This is the town that Juan remembers, which still lives in his heart and mind, but which he had to leave when he was 21. He has Peñarrubia in his soul, and he has dedicated songs to his beloved town:

> I was born in Peñarrubia and I'll remember forever / That pretty little town where I grew up, / Where I wish I could live, but that's impossible. / As you know, or even if you don't, Peñarrubia no longer exists. / In the name of progress, they made a swamp there / And flooded my lovely little town underwater. / I will always remember what they did with you, / Tearing you all up and then demolishing everything. / And as if that were not enough, they sunk you underwater. / I think about you, I will never

forget, and I don't want to forget. (a flamenco-style song written and performed by Juan Mora, 11 October 2016)

Juan was one of the last ones, so he witnessed how the Guadalteba dam's water smothered his town's life. No matter how much it hurt to leave their land, no one could protest, negotiate or complain. However, the Peñarrubia stories attest to how one's roots cannot be removed, and they drive resistance at least to never forget. Ever since they flooded his town, Juan keeps going back to Peñarrubia.

I left on 19 April 1972, and on May 8th I purchased a motorcycle to get to work in Málaga. Then, when I would finish my work on Saturday at noon, when I got home, I would shower, eat, get dressed, get on the motorcycle and come to Peñarrubia and sleep in the field. No one lived in the town anymore. [...] I would sleep in the field, alone, lying there Saturday night, Sunday, and then Sunday night I would head back to Santa Rosalía to get back to work on Monday. [...] I wanted to be back in my environment, on my land. (11 October 2016)

Now his town was an empty space. Each weekend, he would watch the water rise until the reservoir was filled, in May 1979, and his town was buried. Some years ago, Juan purchased a small house on the edge of his drowned town to spend all the time there that he can:

I was working in construction, then I got a job with a very good factory, bottling soft drinks, and I was making good money and not working so hard. But I am not thankful for that, because I wish I could spend my life here and I can't. I have to put up with living far from my land. (11 October 2016)

With the money they received for the expropriation, Peñarrubia's people were forced to move to the colonization towns, mostly Santa Rosalía, and to Catalonia. Land in Santa Rosalía was not as productive as in Peñarrubia. Agustín tells us:

When they fill a reservoir, the water covers the best land, the fields, the townsite – it never covers the poorer land. [...] Santa Rosalía here is not like Peñarrubia. There we had plenty of wheat, fava beans, barley, and chick peas. (13 October 2016)

Another old man, Juan Pozo, tells us with tears in his eyes: 'I still have the keys to my home there. [...] Half of my nights, I dream about Peñarrubia' (13 October 2016). Our interviewees tell us that many of the older neighbours did not survive the dramatic change. Juan Mora recalls: 'elderly folks, accustomed to wandering freely in our town, were stuffed into a big city, buried alive in a flat. They didn't last more than five or six months, and then died, of grief' (11 October 2016).

Industrial and urban growth in Málaga generated employment opportunities for the displaced rural people. Most worked in construction. In Spain, under the hydraulic policy between 1950 and 1970, there are many towns like Peñarrubia that were forcibly expropriated by the state and then evicted by the Guardia Civil police force.

In 2010, Juan Mora Ríos dedicated his book to his town: *Peñarrubia, Two Plazas and Nine Streets*, with verse and prose giving a tour of every corner of his town, bringing the characters in their day-to-day business back to life:

Alongside a pretty river, on a small hill, in the plains surrounded by fig trees, fields, grainfields – there was my town: Peñarrubia was the greatest happiness I had in my life, since my childhood, a lovely Málaga town that no longer exists. (11 October 2016)

The town used to have almost 2000 inhabitants, but now the only family with their permanent residence in Peñarrubia is that of Antonio Escalante and Dolores Morgado. Antonio is a retired watchman who came to watch over the desolated town. They built their little home next to the dam, where their eight children were born, and have stayed there all their lives. Now they welcome visits by the people displaced from Peñarrubia with warm hospitality. 'This town was well-beloved. That is why people come every week to remember their past. Every year they celebrate their romería commemorations, gathering hundreds of families who shared their roots here more than 40 years ago' (Antonio Escalante, 3 February 2017).

Rural-to-urban water transfers, water-use priorities

Nowadays the main use for the Guadalhorce system's water is for the city, Málaga: 1500 l/s continually. The last reservoirs (Guadalhorce and Guadalteba) were supposed to provide 400 litres per person per day, for a population of 500,000 in 1973 (Calvo, 1973). However, the January 2016 census shows the province of Málaga with 1.629 million[1] and a permanent population in the city of 558,287 (CAPMA, 2012), which multiplies during the tourist season. For example, in 2011, during the tourism peak months, the province reached 9 million.[2] From 2000 to 2007, 480,000 urbanization parks were built on the Costa del Sol, and Málaga featured the highest increases in tourist residences along its coastline (CAPMA, 2012). This shows how water from the Guadalhorce Valley is at the service of a floating population and a capitalist tourism model that agglomerates people on the coast.

During droughts, Málaga has priority for city water supply. For example, the mid-1990s' drought completely suspended irrigation and dried up the river completely, eliminating its biodiversity. In the mid-2000s, another extreme drought period, farmers got to irrigate only a few times with a minimal dotation. Meanwhile, the Málaga Municipal Water Company (EMASA) had no restriction on its supply from the reservoirs (van der Kooij, 2011). Spain's normative framework (as in almost all countries) gives priority for domestic water use, so water-supply companies such as EMASA do not have to compete with other usage rights. This norm is on the face of it important and legitimate. However, besides the water categorized and distributed as 'water for domestic use', EMASA also distributes water for 'industrial uses' (including recreational uses); and 'institutional uses' (such as hospital and offices) and green irrigation for public gardens. For example, official data for Málaga show five water parks, a botanical garden, six swimming areas, etc. (CAPMA, 2012). Next, 'domestic water' also fills swimming pools, waters private gardens, beach showers, washing cars and streets, etc. (e.g. Hazeleger & Boelens, 2003). Since 1989, in view of high urban consumption, especially during summer, EMASA has diversified its water supply sources. Nevertheless, this urban demand continues to extract most of its water from the Guadalhorce canal (CAPMA, 2012).

Hydro-territorial transformations in the Guadalhorce Valley described above reveal profound contradictions in water and economic policy. In the 1960s, this zone had plans for agricultural production as well as tourism in Málaga. Then, in the 1990s, the real estate boom built rural homes for tourists in the valley, enabling north-western European elites to settle in the most productive zones, demanding irrigation water for their own recreational purposes (Hazeleger & Boelens, 2003).

Ironically, the 'colonization towns' located *in zones without irrigation* became the buffer zones for migration by people whose lands were seized for hydraulic projects in the

Guadalhorce Valley. At this time, massive immigration by capitalist residential tourism is happening *inside the irrigation system*, funded by and therefore taking advantage of Spain's public monies. Further, these outsiders usually fence in their private farms, isolate themselves from community life, ignoring all irrigation customs and scheduling, thereby breaking down the last remnants of local water collectivity and culture.

Another very controversial political and social issue are the other totally elite water uses in Guadalhorce: the urbanization model based on golf courses (irrigation water and drinking water uses) (see also Villar Lama, 2013). The water system for Serranía de Ronda, in Guadalhorce, shows 76 golf courses with 56 clubs and, in the study area of influence, eight golf courses with five clubs (CAPMA, 2012, p. 109). Authorities estimate that the entire system consumes 22.3 hm^3 annually (14.68 hm^3 pumped from underground; 0.28 hm^3 from surface water; and 7.34 hm^3 from wastewater). Although since the end of 1980, Spanish law requires the watering of golf courses with treated wastewater (Espejo Marín y Cànoves, 2011), this is not enforced. Rural people say that the aquifer is affected and irrigation scheduling restricted during droughts partly to serve this urban elite activity. This situation deepens when territorial and urban legislation encouraged the possibilities to expand golf courses in rural areas of the valley as projects of 'tourist interest'.[3]

Here we see a change from governed-based disciplinary territorial transformation towards market-led neoliberal governmentality. Territorial planning runs wild – or, rather, is 'governmentalized' – according to market forces and incentives, powerful economic interests and urban real estate speculation. This knits together a network of interests: real estate developers, foreign investors, speculators and government agencies.

This means that water demand for Málaga's city water supply has drastically altered territoriality in the Guadalhorce Valley. First, historically, by eliminating community systems and local water self-governance during Franco's times, putting them all into an integrated, uniform, top-down system that actually shifted more water to where power was concentrated, the city of Málaga, at the expense of irrigation. Further, history shows that much bureaucratized irrigation water flowed not to poor small farmers but to large landowners; and now is reoriented toward elite residential tourism. Second, by building water mega-works designed to suit the city's interests, rural communities and families were uprooted and displaced from their territories. As an alternative livelihood, they were governmentalized into 'colonization towns. Third, irrigation water is continually reduced, to be transferred to the unbridled growth of urban and tourist industry. Del Moral (2009) explains that the prevailing urban development dynamics based on building scattered homes, under used with the predominant seasonal sun-and-beach model, means heavy consumption of water and territory. Meanwhile, groundwater extraction is increasing without any local and social regulation, meaning that rural tensions, also among irrigators, are on the rise.

Reflections and conclusions

By studying the history of (technocratic, regenerationist-progressive) hydro-territorial imaginaries and transformations in the Guadalhorce Valley, this paper has shown how materializing regenerationist hydraulic utopia during the past century has turned into a dystopia for many of the valley's rural families. Seeking a radical and simultaneous transformation of society and nature (people and land); expanding waterworks; nationalizing and spatially reorganizing water supply under expert rule; and bringing more land under irrigation, the regenerationist hydraulic utopia aimed to bring shared benefits for all social classes. However, the imagined development through modernity's splendid benevolence and enlightened engineers led by an iron-fisted surgeon concealed the unimagined human and ecological disasters.

As a legacy of Joaquín Costa, and under the political leadership of regenerationists as Rafael Benjumea, the Corps of Engineers arose as the beacon for development in Spain and in the Guadalhorce Valley. Expert, positivistic knowledge, implemented through objectifying science, would have the power to 'colonize the country inwardly', lumping together local hydro-social territories to consolidate national water identity, controlling water distribution through river basin confederations. This meant that each dam was a governmental mechanism to discipline and standardize society–nature relations in each river basin.

Notwithstanding the regenerationists' hydro-social transformation dreams, as we have shown, the hydraulic utopia materialized and expanded only under Francoist violent governmental strategies. Forced mobilization of Guadalhorce's rural communities to 'colonization towns' heavily influenced and 'corrected' the valley's rural people's livelihoods and practices. The fear imposed during the Franco regime was the mechanism to induce obedience among the people, normalizing the damming and displacement of their waters and ideologies. Even during the transition to democracy, the cornerstones of hydraulic policy remained unquestioned. The monolithic foundations of productivity, neutrality and expert knowledge kept rural peoples' dystopia invisible.

This forced rural dwellers in the Guadalhorce Valley to see and feel the other side of the hydraulic utopia: of demolishing Arab waterworks, losing self-governance of their water systems, water grabbing first by large landholders and now by tourism when summers are dry. Hydraulic utopia 'in the national interest' meant accepting as normal that rural families had to sacrifice everything they had in exchange for the 'well-being for the majorities'. This emptied rural territories and played into the hands of energy and water demand for tourism and other urban power concentrated in Málaga.

Guadalhorce is now organized chaotically, subject to the whims of market laws and the dynamics of urban development speculation – neoliberal governmentality. Urban and tourist industry growth in Málaga is individualizing land and water management in the irrigation system, and reinforcing water transfers from Guadalhorce communities to the city. This configures a new hydro-territorial order in the valley in which water flows to suit urban and elite needs.

Even so, with their past underwater and their Guadalhorce transformed, the voices of resistance persist, through displaced neighbours' feelings of rootedness, keeping their memories alive, bridging the present with the past and future, encouraging and re-signifying their hydro-social territory. Unfolding memory has meant liberation from the dystopias contained in this hydraulic utopia that swallowed their towns and walled in their river. Now their memories urge them back to the countryside, building alliances of solidarity and reviving their intimate linkages with the river and its lands. This way, they aim to revive and truly regenerate the river and peoples of Guadalhorce.

Notes

1. See http://www.laopiniondemalaga.es/malaga/2016/01/21/malaga-gana-poblacion-ano-pese/823083.html.

2. See http://www.elmundo.es/elmundo/2012/03/08/andalucia_malaga/1331216401.html; and http://www.juntadeandalucia.es/institutodeestadisticaycartografia/atlashistoriaecon/atlas_cap_56.html.
3. https://www.juntadeandalucia.es/ciudadania/procedimientos?p_p_id = catalogoProcedimientos_WAR_catalogoProcedimientosportlet&p_p_lifecycle = 0&p_p_col_id = column-2&p_p_col_count = 1&_catalogoProcedimientos_WAR_catalogoProcedimientosportlet_idProcedimiento = 1506&_catalogoProcedimientos_WAR_catalogoProcedimientosportlet_cmd = detalle.

Acknowledgments

We thank the International WaTERS Network (Water Related Training Education and Research in the Global South) and the Justicia Hidrica / Water Justice Alliance, in whose activities this research is embedded.

Disclosure statement

No potential conflict of interest was reported by the authors.

Funding

This was was supported by an International WaTERS research and training network SSHRC grant 890-2013-0020.

ORCID

Bibiana Duarte-Abadía http://orcid.org/0000-0003-4203-050X
Rutgerd Boelens http://orcid.org/0000-0002-8412-3109

References

Achterhuis, H. (1998). *De erfenis van de utopie* [Utopia's Heritage]. Amsterdam: Ambo.

Achterhuis, H., Boelens, R., & Zwarteveen, M. (2010). Water property relations and modern policy regimes: Neoliberal utopia and the disempowerment of collective action. In R. Boelens, D. Getches, & A. Guevara (Eds.), *Out of the mainstream: Water rights, politics and identity* (pp. 27–56). London: Earthscan.

Acosta Bono, G., Gutierrez Molina, J., Martinez Macias, L., & Del Rio Sanchez, A. (2004). *El canal de los presos (1940–1962). Trabajos forzados: De la represión política a la explotación económica*. Barcelona: Critica.

Agencia Estatal (1961). *Plan Coordinado de Obras de la Zona Regable del Río Guadalhorce (Málaga)*. Boletín Oficial. Retrieved from https://www.boe.es/boe/dias/1961/03/28/pdfs/A04769-4773.pdf

Baviskar, A. (2007). *Waterscapes: The cultural politics of a natural resource*. Delhi: Permanent Black.

Boelens, R. (2017). *Rivers of Scarcity. Utopian water regimes and flows against the current*. Wageningen: Wageningen University. http://edepot.wur.nl/432727

Boelens, R., Hoogesteger, J., Swyngedouw, E., Vos, J., & Wester, P. (2016). Hydrosocial territories: A political ecology perspective. *Water International*, *41*(1), 1–14. doi:10.1080/02508060.2016.1134898

Boelens, R., & Post Uiterweer, N. (2013). Hydraulic heroes: The ironies of utopian hydraulism and its politics of autonomy in the Guadalhorce Valley, Spain. *Journal of Historical Geography*, *44*, 44–58. doi:10.1016/j.jhg.2012.12.005

Brotons, J., & García, C. (1999). *El Embalse del Chorro, un hito en la política hidráulica en el umbral del siglo XX*. Málaga: Comisión Hidrográfica del Sur.

Calvo, J. M. (1973). Resumen del Plan Coordinado del Guadalhorce. *Revista Jabega, 1*, 31–34.
Camprubí, L. (2013). *Engineers and the making of the Francoist regime*. Cambridge MA: MIT Press.
Cantero, O. (1995). El Plan General de Canales de Riego y Pantanos de 1902. In A. G. Olcina & A. M. Gil (Eds.), *Planificación Hidráulica en España* (pp. 107–136). España: Caja de ahorros del mediterráneo.
Closas, A. (2018). Groundwater, the state, and the creation of irrigation communities in Llanos del Caudillo, Spain. *Water Alternatives, 11*(1), 19–39.
Consejería de Agricultura, Pesca y Medio Ambiente (CAPMA) (2012). *Plan Hidrológico de la Demarcación Hidrográfica de las Cuencas Mediterráneas Andaluzas*. Retrieved from www.juntadeandalucia.es/medioambiente/site/
Costa, J. (1911). *Política hidráulica: Misión social de los riegos en España* [Water policy: social mission of irrigation in Spain]. (Vol. 2). Madrid: Biblioteca J. Costa.
Costa, J. (1967). *Oligarquía y caciquismo. Colectivismo agrario y otros escritos*. Madrid: Alianza.
de Andalucía, J. (2008). *Propuesta de actuación en la zona regable del Guadalhorce*. Málaga: Junta de Andalucía.
Del Moral, L. (2009). Nuevas Tendencias en Gestión del Agua, Ordenación del Territorio e Integración de Políticas Sectoriales. *Revista Electrónica de Geografía y Ciencias Sociales, 18*, 741–798.
Del Moral, L., & Saurí, D. (1999). Changing course: Water policy in Spain. *Environment Science and Policy for Sustainable Development, 41*(6), 12–36. doi:10.1080/00139159909604640
Delgado, M., & Del Moral, L. (2016). *Los megaproyectos en Andalucía. Relaciones de poder y apropiación de riqueza*. Aconcagua: Sevilla.
Duarte-Abadía, B., Boelens, R., Du, P.L. (2019). Mobilizing water actors and bodies of knowledge. The multi-scalar movement against the Río Grande Dam in Málaga, Spain. *Water* 2019, *11* (3), online: doi:10.3390/w11030410
Espejo Marín, C., & Cànoves, G. (2011). Política de usos del agua en los campos de golf en España. *Documents d'anàlisi geogràfica, 57*(2), 255–277.
Fernández Clemente, E. (2000). *De la utopía de Joaquín Costa a la intervención del estado: Un siglo de obras hidráulicas en España*. Zaragoza: Universidad de Zaragoza.
Fletcher, R. (2010). Neoliberal environmentality: Towards a poststructuralist political ecology of the conservation debate. *Conservation and Society, 8*(3), 171. doi:10.4103/0972-4923.73806
Foucault, M. (1991[1978]). Governmentality. In G. Burchell, C. Gordon, & P. Miller (Eds.), *The Foucault effect: Studies in governmentality* (pp. 87–104). Chicago: University of Chicago Press.
Frutos Mejías, L. M. (1995). Las confederaciones sindicales hidrográficas (1926–1931). In G. Olcina & A. Morales Gil (Eds.), *Planificación Hidráulica en España* (pp. 181–256). España: Caja de ahorros del mediterráneo.
Gómez Mendoza, J. (1992). Regeneracionismo y regadíos. In A. Gil Olcina & A. Morales Gil (Eds.), *Hitos históricos de los regadíos Españoles* (pp. 231–262). Secretaría General Técnica. Madrid: Ministerio de Agricultura. Pesca y Alimentación.
Gray, J. (2007). *Black mass. Apocalyptic religion and the death of Utopia*. London: Allen Lane.
Harris, L. M. (2002). Water and conflict geographies of the Southeastern Anatolia Project. *Society &Natural Resources, 15*(8), 743–759. doi:10.1080/08941920290069326
Hazeleger, B., & Boelens, R. (2003). A place in the sun. *Documentary on Guadalhorce Valley*. Wageningen: Agrapen and Wageningen University.
Hommes, L., & Boelens, R. (2017). Urbanizing rural waters: Rural–urban water transfers and the reconfiguration of hydrosocial territories in lima. *Political Geography, 57*, 71–80. doi:10.1016/j.polgeo.2016.12.002
Hommes, L., & Boelens, R. (2018). From natural flow to 'working river': Hydropower development, modernity and socio-territorial transformations in Lima's Rímac watershed. *Journal of Historical Geography, 62*, 85–95. doi:10.1016/j.jhg.2018.04.001
Hommes, L., Boelens, R., Harris, L., & Veldwisch, G. J. (2019). Rural–urban water struggles: urbanizing hydrosocial territories and evolving connections, discourses and identities. *Water International, 44* (2), 81–94. doi:10.1080/02508060.2019.1583311
Kaika, M. (2006). Dams as symbols of modernization: The urbanization of nature between geographical imagination and materiality. *Annals of the Association of American Geographers, 96*, 276–301. doi:10.1111/j.1467-8306.2006.00478.x

Kumar K. (1987). *Utopia and anti-utopia in modern times.* Oxford: Blackwell.

Lafuente, I. (2002). *Esclavos por la patria: La explotación de presos bajo el Franquismo.* Madrid: Ediciones Temas de Hoy.

Lara, S. (2002). *100 años Del Chorro.* Alora: Centro de Ediciones de la Diputación de Málaga.

Levitas, R. (1990). *The concept of Utopia.* London: Allan.

Linton, J., & Budds, J. (2014). The hydro-social cycle: Defining and mobilizing a relational-dialectical approach to water. *Geoforum, 57,* 170–180.

Lopez-Gunn, E. (2009). *Agua para Todos: A new regionalist hydraulic paradigm in Spain.* Water Alternatives, 2/3, 370–94.

Martín-Gaite, C. (2003). *El Conde del Guadalhorce. Su época y su labor.* Madrid: Ediciones Turner.

Maurice, J., & Serrano, C. (1977). *J. Costa: Crisis de la restauración y populismo (1875–1911).* Madrid: Siglo XXI Editores.

Marks, D. (2019). Assembling the 2011 Thailand floods: Protecting farmers and inundating high-value industrial estates in a fragmented hydro-social territory. *Political Geography,* 68, 66–76. doi:10.1016/j.polgeo.2018.10.002

Meehan, K. (2013). Disciplining de facto development: Water theft and hydrosocial order in Tijuana. *Environment and Planning D: Society and Space* 31, 319–336. doi:10.1068/d20610

Mülberger, A., Vilaró, M., Tirado, F., & Domènech, M. (2007). *Historia, política y ciencia: El papel de los expertos en el debate sobre el agua en España.* Retrieved from http://www.recercat.net/bitstream/2072/4783/1/Recerca+ historia+ aigua+Espanya.pdf.

Ortí, A. (1984). Política hidráulica y cuestión social: Orígenes, etapas y significados del regeneracionismo de Joaquín Costa. *Agricultura y Sociedad, 32,* 11–107.

Roberts, C. R. (2002). Drought management in the Río Guadalhorce region of Andalucía, Southern Spain. *Land Degradation & Development, 13,* 151–163.

Sanchis-Ibor, C. (2012). La confederación fallida. Administracion y usuarios en la cuenca del Turia (1928–1936). *Cuadernos de Geografía, 91–92,* 19–42.

Seemann, M. (2016). Inclusive recognition politics and the struggle over hydrosocial territories in two Bolivian highland communities. *Water International, 41*(1), 157–172. doi:10.1080/02508060.2016.1108384

Swyngedouw, E. (1999). Modernity and hybridity: Nature, regeneracionismo, and the production of the Spanish waterscape, 1890–1930. *Annals of the Association of American Geographers, 89*(3), 443–465. doi:10.1111/0004-5608.00157

Swyngedouw, E. (2007). Technonatural revolutions: The scalar politics of Franco's hydro-social dream for Spain, 1939–1975. *Trans Inst Br Geogr, 32,* 9–28.

Swyngedouw, E. (2015). *Liquid power: Contested hydro-modernities in twentieth-century Spain.* Cambridge MA: MIT Press.

Swyngedouw, E., & Boelens, R. (2018). '.... Not a single injustice remains'. Hydro-territorial colonization and techno-political transformation in Spain. In R. Boelens, T. Perrault, & J. Vos (Eds.), *Water Justice* (pp. 115–133). Cambridge: Cambridge University Press. doi:10.1017/9781316831847.008

Swyngedouw, E., & Williams, J. (2016). From Spain's hydro-deadlock to the desalination fix. *Water International, 41*(1), 54–73. doi:10.1080/02508060.2016.1107705

van der Kooij, S. (2011). *Guadalhorce Watershed. Justicia hidrica/water justice report.* Wageningen: Wageningen University.

Villar Lama, A. (2013). La mercantilización del paisaje litoral del Mediterráneo Andaluz: El caso paradigmático de la Costa del Sol y los campos de golf. *Revista de Estudios Regionales, 96,* 15–30.

Vos, J., & Boelens, R. (2018). Neoliberal water governmentalities, virtual water trade, and contestations. In R. Boelens, T. Perrault, & J. Vos (Eds.), *Water Justice* (pp. 283–301). Cambridge: Cambridge University Press. doi:10.1017/9781316831847.019

OPEN ACCESS

The political construction and fixing of water overabundance: rural–urban flood-risk politics in coastal Ecuador

Juan Pablo Hidalgo-Bastidas and Rutgerd Boelens

ABSTRACT

Ecuador's mega-dam project aims to control Chone city's flooding hazards, but it submerges peasants' territories – legitimized by 'modern city/majority benefit' versus 'rural backward/sacrifice-able minority' discourse. Presented as disordered, unruly and needing domestication, peasants must follow urban imaginaries and safeguard modern-urban progress. Policy-makers' water overabundance discourse presents 'flood risk' as a natural and techno-managerial problem, hiding how unequal power balances establish 'high-value' (urban/elite) areas as protection zones and rural areas as sacrifice zones. Excessive water is stored in rural areas, neglecting peasants' livelihoods and governance forms. The paper's political ecology approach displays the 'water overabundance' discourse as a techno-political, naturalized construct that profoundly impacts rural–urban hydro-territoriality.

Introduction

In the thematic fields of rural–urban water governance, critical scholarship has identified the theme of water transfers from rural to urban areas for urban water supply as a prominent issue. Such transfers are commonly motivated by a powerful water-scarcity discourse, as noted by, among others, Bakker (2010), Hommes and Boelens (2017), Lynch (2013), Riaz (2002) and Swyngedouw (2004). This discourse on a city's water scarcity commonly sidelines intra-urban leakages, exclusions and mismanagement to justify bringing water that is presumably unused from apparently water-abundant and pristine rural water reserves. Water-scarcity discourse is criticized for legitimizing and obscuring the contentious ways in which water resources are taken from these rural territories to benefit urban areas. What is likewise important but rarely addressed is how *water-overabundance* discourse similarly serves to support and legitimize the adoption of depoliticized urban-biased water policy models (cf. Alatout, 2009). This becomes prominent, for instance, in the case of possible flood risks for cities. There are similar shifts and movements of water under notions of water overabundance. Though given far less critical attention, in the same way water overabundance tends to neglect rural livelihoods. Where is the overabundant water kept? What transformations does this bring to rural and urban areas? How does the nexus between modern water technologies (e.g., dams) and flood risk work?

In the past century across the globe, urban centres have become consolidated as privileged settings for modern development and societal progress, even as they remain deeply stratified (Banister & Widdifield, 2014; Baviskar, 2007). Precisely because cities gather large numbers of stakeholders and diverse economic activities, among their most pressing problems is the issue of water access and control. The growing uncertainty about the potential effects of climate change (e.g., droughts and flooding) and permanent, often unplanned, migration from the countryside to the city are placing even greater pressure on urban water governance (Molle & Berkoff, 2006). In many cities, these problems involve not only water supply but also keeping overabundant water away. In this context, hydraulic mega-infrastructure projects such as inter-basin transfers and multipurpose dams are widely accepted among politicians, public planners, financing agencies and private-sector stakeholders such as construction companies. Such projects facilitate the redistribution of water to the strongest actors, whereby in many cases large cities acquire a prominent role. Rural areas are commonly underestimated and framed as the *other*. For instance, Duarte-Abadía and Boelens (2016, 2019, in this issue) show how, in the context of rural–urban water conflicts in Colombia and Spain, peasant water cultures and knowledge frameworks are presented as backward compared with those of modernist urban-biased water provision projects (see also Roa-García, 2014; Romano, 2017; and Hommes, Boelens, Harris, & Veldwisch, 2019, in this issue). Johnston (2018) and Lynch (2019) present the illustrative case of Chixoy in Guatemala, where rural areas were written off as empty-able (see also, e.g., Baviskar, 2007; and Nixon, 2010). Alternatively, as aquifer recharge areas and other bodies of fresh water originate precisely in rural areas, they are often presented as places to be conserved, or depicted as the places where the *natural* capacity for flood mitigation lies. This interaction creates and aggravates tensions between urban and rural water governance.

As will be argued throughout this paper on Ecuadorian water development, dominant modernist water policies and models are justified by categorizing rural territories as areas that can be sacrificed for urban flood protection implicitly or explicitly portraying urban areas as developed while depicting rural territories as underdeveloped and backward (see, for instance, Radcliffe & Westwood, 2005, for an analysis of Ecuador's modernist cultural politics). This argument is illustrated through the discourses and practices deployed to legitimize the urgent call to prevent floods (water overflows) affecting the city of Chone in Ecuador (Figure 1). In analyzing these flows and infrastructural efforts, we offer scrutiny of ways that the government deploys constructed hierarchies of urban over rural. Using a depoliticized water overabundance discourse, the government also justified the construction of the Chone multipurpose mega-dam and the associated transformation of rural surroundings.

This contribution is based on historical and ethnographic research carried out in the city of Chone and rural Río Grande communities located in the Chone multipurpose mega-dam's influence area (Figure 1). The field research was performed through

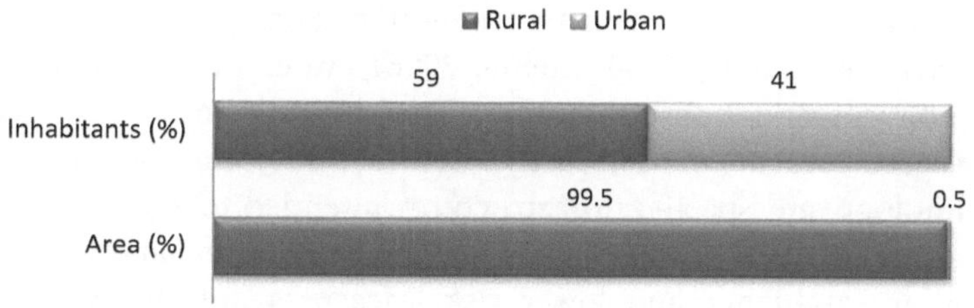

Figure 1. Chone's dam and the main rivers flowing through the city of Chone.
Source: Authors.

extensive field visits between 2014 and 2016 and the first author's professional experience within the project from 2010 to 2011. Participatory observation, semi-structured interviews, a literature review and secondary sources (historical archives, newspaper articles, official reports) were the main data-collection methods.

Water overabundance and the urban–rural water nexus

Water governance is socially constructed and politically informed. It involves, for instance, divergent interests and contradictions regarding water control and access in cities (e.g., Ioris, 2016; Swyngedouw, 1995), in rural sectors (Hidalgo, Boelens, & Vos, 2017; Mena-Vásconez, Boelens, & Vos, 2016) and the relations between them (e.g., Linton & Budds, 2014; Molle & Berkoff, 2006; Riaz, 2002). The different means (public policy, laws, governance models, infrastructure/technology, etc.) used to appropriate, accumulate and/or distribute water are deeply influenced by different notions of how to understand the relationships between nature, technology and society (Baghel, 2014; Swyngedouw, 2004). In the discussions regarding mega-hydraulic infrastructure and large dams, proponents typically construct water discourses around two 'naturalized' extreme conditions or threats to legitimize their positions: water scarcity presented as droughts and water overabundance deployed as floods (cf. McCully, 2001).

The discourse created around physical water scarcity as the basis to develop 'good water governance' is widely promoted, but also criticized. Goldman (2007) and Lynch (2013), for example, show how the discourse of scarcity is historically promoted by international institutions such as the United Nations (UN), the Food and Agriculture Organization (FAO) and the World Bank (WB) and adopted by governments to justify their actions regarding 'good water governance'. Lynch insists that 'the policy prescriptions of international institutions seeking to transform fresh water governance [...] are driven by fear of an impending crisis due to an *absolute scarcity* of fresh water' (Lynch, 2013, p. 70, added emphasis). Through the Peruvian case, Lynch shows how this universalized discourse tends to present water governance as an apolitical, natural matter, while eclipsing essential aspects of its social, political and economic construction. Thus, water scarcity is referred to only as a natural problem caused by nature and climate change, rather than as a problem of distribution or of uneven power relations (cf. Bakker, 2010; Hommes & Boelens, 2017; Swyngedouw, 2004).

As with the case of the social construction of water scarcity (e.g., Aguilera-Klink, Pérez-Moriana, & Sánchez-García, 2000; Mehta, 2006), water overabundance is similarly a relational concept and political notion. Thereby, the social construction of abundance is translated into *flood risk*, partly used to justify the implementation of modern water technology and specific infrastructures intended to protect cities as key sites of progress (e.g., mega-dams). With such a logic, state planning and water agencies define which areas are at higher and lower risk. Measures are chosen to decrease potential flooding contingencies, privileging certain areas over others (e.g., Porter & Demeritt, 2012; Roth & Warner, 2007). When defining these areas, there is usually a correlation between increasing risk and economic assessment: the higher the economic value, the higher the risk (Warner, 2010).

From this perspective, mega-hydraulic projects are at times deployed as the panacea to control floods. These infrastructures are generally located upriver from urban centres and/or rural areas with high levels of capital investments (e.g., Warner, Hoogesteger, & Hidalgo-Bastidas, 2017) and politically powerful interest groups (Marks, 2019; Roth et al., 2018). As shown by Osti (2017), the focus of implementing actions and infrastructure to prevent flood risk is contentious and involves the need to find 'rural spaces to occupy [...] the image of security, and the search for an urban–rural cost–benefit balance' (p. 266). This haggling between urban and rural suggests that – as in the case of scarcity – so-called overabundance is not always merely physical and natural. As a techno-political and discursive policy notion, flooding risk may also become a contingency swayed by profoundly social, economic and political reasons, and power relationships. This also means that, in utilitarian terms, certain territories imagined as *less valuable* could be *sacrifice-able* in the name of common well-being (Nixon, 2010), privileging the so-called majority who live in *higher value* areas. Thus, the social and political construction of 'water overabundance' is also a powerful driver transforming policy, institutions, water infrastructure, and everyday people's lives and territories in rural–urban contexts (e.g., Hommes, Boelens, Duarte-Abadía, Hidalgo-Bastidas, & Hoogesteger, 2018). In sum, water overabundance relates to a hydro-political notion that in political–economic practice often becomes technified, universalized and naturalized. What is considered as water-overabundant is a social and a political construction (cf. Alatout, 2009; Erensu, 2013). Through the specific case of Chone, this paper helps one to understand how water-overabundance discourse and management practice are defined by sociopolitical and economic power relationships and how they serve to render invisible rural–urban water inequities.

Therefore, beyond viewing urban or rural separately, the study of the urban–rural dialectic is important, requiring a political ecology focus of urbanization/urban–rural territorialization. The urban–rural or country–city relationship is not one way; urban and rural regions deeply influence each other and are fundamentally interlinked. What we do make explicit is that the linkage between these two spaces often is inequitable and driven by power relationships that have historically positioned urban centres as the referents for development, while rural areas have been relegated as places that can be sacrificed (Boelens, 2015a; Nixon, 2010). In this representation, 'underdeveloped' rural spaces may become 'developed' through adopting (e.g., urban) dominant norms and values (Shah, Liebrand, Vos, Veldwisch, & Boelens, 2018; Hommes et al., 2019, in this issue). Since colonial times, in Ecuador as in other

Andean countries, the countryside usually has been on a tightrope between being labelled as *different* and, at the same time, as *potentially equal* (Boelens, 2015b). It is constructed as different because of its 'backwardness' compared with the city, but with potential to catch up and become equal by adopting and adapting to the models, values, and identities imagined and imposed by those who are in power (cf. Radcliffe & Westwood, 2005; Hommes et al., 2019, in this issue; Torio, Harris, & Angeles, 2019, in this issue). Thus, our understanding of water governance, in any context, goes beyond the idea of humans governing water but captures, in particular, the notion of 'governing people through water, as ways of organizing power and decision-making to bring about environmental control and societal order at once' (Boelens, 2015a, p. 13).

The return of the state: the citizens' revolution – good living and modernity

In 2007, a new government came to power in Ecuador. Its battle cry was to re-establish the nation with a revolution by the citizenry. This revolution would leave behind what then-President Rafael Correa framed as 'the long, woeful neoliberal night'. The overarching goal of his political movement, Patria Altiva i Soberana (PAIS – Proud, Sovereign Nation), was to construct Socialism for Good Living. Guided by this premise wielding a progressive discourse of social justice, equality and sovereignty, the government of the so-called Citizens' Revolution made an array of efforts to strengthen the state's role and presence, which had been dismantled by neoliberalism and its consequent social, political and economic crises since the 1990s. Part of the itinerary for this dawning of a new age was to profoundly modernize the state, recover the torn-down institutions and bring development expressed as *Buen Vivir* (Good Living) to every corner of the nation. The most visible efforts to achieve this included their alliance with other political forces (indigenous movements, critical scholars, rural people, labour unions, etc.) and citizen forces to construct, in 2008, a new constitution guaranteeing rights, as never before. This new constitution emphasizes the importance of citizens' participation, recognition of pluri-nationality, ample introduction of access to water as a human right and respect for the rights of nature (e.g., Boelens, Hoogesteger, & Baud, 2015; Hidalgo et al., 2017; Radcliffe, 2012).

Although they began with quite progressive, inclusive proposals, over the years this political movement has received criticism even from many groups that initially backed it. This criticism has been based mainly on contradictions between governmental discourse and practice (e.g., Acosta et al., 2013; Gudynas & Acosta, 2010). Most criticism involves the government's inability to include stakeholders who think differently, the criminalization of protests, authoritarianism, lack of independence for branches of the state, de-legitimization and consequent weakening of autonomous societal organization (Acosta et al., 2013). Part of the criticism included the government's uses of the concept of *Buen Vivir* – social organizations and other critical social actors (e.g., indigenous movement) accused the government of having captured the concept to make it instrumental to suit its own purposes, including particular economic and political interests (Van Teijlingen & Hogenboom, 2016). While, on the one hand, government institutions have embraced a combination of discourses of progress, well-being and efficiency, presented as the scientific–bureaucratic truth, on the other hand, dissident voices inside and outside the state have been demanding recognition for alternative modes of thinking and acting regarding societal development. This made the government disqualify opponents in a profound manner: portraying them as against development, backward and even pre-modern. This tension and criticism surfaces particularly in state interventions that conveniently ignore and infringe citizens' rights that are consecrated in its own constitution (Radcliffe, 2012; Valladares & Boelens, 2017; Van Teijlingen, 2016).

One core thrust of the government's political project was to change the production and energy matrix, for which technocratic water management has played a leading role (Hidalgo-Bastidas, Boelens, & Isch, 2018; Hoogesteger, Boelens, & Baud, 2016). As part of the new institutional arrangements to promote this approach, and for managing water and other resources defined as strategic, the Ministry of Strategic Sectors (MICSE) was created, with the National Water Secretariat (SENAGUA) and the Strategic Ecuador Public Enterprise (EEEP). The MICSE governs policy to manage strategic resources, the SENAGUA issues national policy on water management and builds major water projects, and the EEEP provides economic compensation to communities affected by construction and implementation of strategic national projects, with funding from petroleum and mining revenues. Despite the progressive, participatory discourse, megaproject design, construction and implementation have faced resistance from affected local communities. Surprisingly for a government with these leanings, such projects have been tinged with state violence, forced evictions, scanty participation in decision-making, ignored prior consultation and questionable environmental impact studies.

In this context, the state's dominant stereotype of *Buen Vivir* delegitimizes any other forms of development and territoriality, especially those opposed to the works and actions of the official political project (Van Teijlingen, 2016). As then-President Correa put it in his speech, inaugurating a dam built by his government:

> Special greetings for the families affected by the reservoir, who have collaborated patriotically, knowing that this was for the common well-being. They made a sacrifice […], they made an effort, and I think that their efforts have been fairly compensated; they have been resettled, and now they have a decent home, made of cement. (presidential speech, Chone, 27 June 2013)

He implicitly suggests that, thanks to people's alignment with the official development project, they have a *civilized* home, unlike the one they used to have, which was not decent. However, later in that same speech, he complains about the inhabitants of Patricia Pilar, a community that opposed the dam construction somewhere else (Hidalgo-Bastidas & Boelens, 2019):

> Certain groups were opposed […] and we redesigned the project. […] The reservoir is a thousand hectares; the original was to be three thousand hectares, but that would have flooded Patricia Pilar. Patricia Pilar has about 2.000 families, we the Citizens' Revolution could have given them extraordinary houses – like the families who were relocated now – to make a new town, with basic utilities, planning, etc. […], but they showed a lack of vision, opposition to any change, defending their own turf rather than seeing the common good – which will make the world worse rather than better.

This way, people and local territories who aligned with the official project's models and guidelines were portrayed as *radically different*, backward rebels, while they could have become *potentially equal*, belonging to the so-called majority.

The city will not be flooded anymore! the Chone multipurpose dam and urban–rural water paradoxes

The city of Chone is located on the Ecuadorian coast and is the capital of the canton of the same name.[1] This medium-sized city occupies 0.5% of the canton's area, with nearly 52,000 inhabitants. Contrary to what the government discourse depicted, most of the population is scattered throughout the rural areas which account for over 99% of the canton's area (Figure 2). Further, about 70% of the population depends on agriculture or livestock, consolidating the rural area as the canton's and city's most important socioeconomic base of subsistence.

One particular characteristic of this city is that it is located at the bottom of the valley, along the banks of the Chone River (Figure 1). This means that, during each rainy season (from December to May), the city floods when the river and its main tributaries – the Grande (Río Grande), Garrapata and Mosquito rivers – overflow. For this reason, since the early 1980s, infrastructure alternatives have been planned to control flooding of the urban zone. Along with a drainage canal, a mega-dam on the Grande was built as part of the multipurpose project inaugurated by ex-President

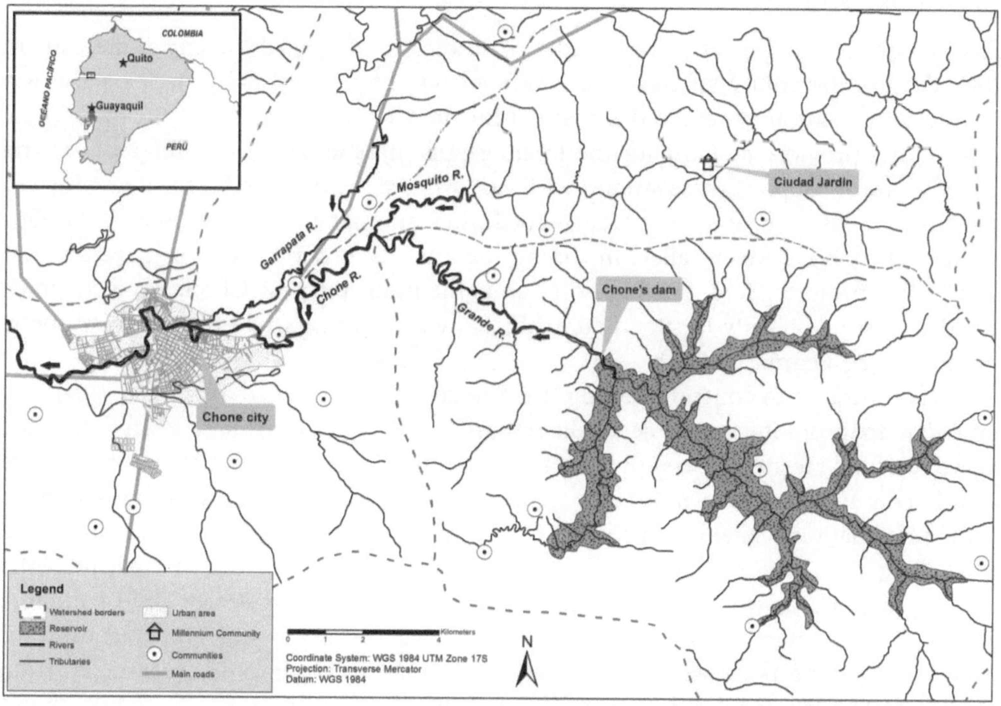

Figure 2. Comparison between rural and urban in Chone's canton.
Source: authors' elaboration.

Correa in November 2015 (Figure 1). The main infrastructure of the Chone multi-purpose system is the dam, 59.5 m high, over the Río Grande.

The project was promoted for its multiple purposes apart from flood control, including provision of irrigation for lower zones (around 2200 ha) as well as provision of domestic water supply for the city.

Chone dam's justification: emphasis on urban risk and development

Despite the importance of rural livelihoods, from the very first discussions about the project, the city of Chone was depicted as a privileged space vis-à-vis the rural areas. Part of this positioning involved reinforcing the discourse about the necessity to address urban flood risk and development centring on the city's image. While flooding risk was the main justification for the dam project, the development discourse played a key role, above all in creating an image of *the other*, of rural life as less developed or backward (cf. Boelens, 2015b; Radcliffe & Westwood, 2005). Neglecting the rural problems of water overabundance, for the first time in November 1979, the largest-circulation Ecuadorian newspaper presented flooding of the city as the main issue:

> Requests from Chone canton will be answered: An expanded meeting will be held in the CRM Manabí Rehabilitation Center [regional entity in charge of planning and managing the Province's water resources up to 2009]. The Governor of the province stated that deviating the Chone River was the most important project because, if the solution of this issue is delayed, this city could be flooded again next rainy season. (*El Comercio*, 26 November 1979)

This urgency led to almost three decades of studies to safeguard the city's interests. In 1983, the first feasibility studies recommended building a regulating dam on the Río Grande, an irrigation system and a channel for the Chone River around the city. These early studies provide the basis for the final designs in 1986. For economic reasons and lack of political support, these studies were not implemented immediately. The project's definitive reactivation was in February 2008 after another heavy rainy season flooded much of the Chone River valley, including the city. Ex-President Correa gave a major push to this project: 'Here, the final solution is the multi-purpose Chone project, which would be justified if only for flood control […], we are going to do this project!' (press conference, 24 February 2008).

It was so urgent to control flooding that the government waged a major propaganda campaign to promote the project. Gigantograph signage all around town reminded everyone of the recent histories of the city's flooding. Complaints and arguments by Río Grande communities were indirectly or tacitly disregarded. Rumours and gossip were all the information available to people living in the rural area where the dam site would be implemented. Farmers, however, presented a calculation showing that permanently flooding their farms would cost the city and province losses of nearly US$14 million. However, the farmers' calculation was ignored, 'because it was not prepared by technicians and engineers, as were the official calculations', according to a technician from the SENAGUA (personal communication, 1 September 2015).

The prioritization of the city over rural areas as a main justification for the project was reflected in the environmental impact assessment (EIA) and the social participation process, which, furthermore, did not follow the established procedures. As an affected farmer put it, 'their point, SENAGUA's, is to make the reservoir here, and then see where they can relocate us [...] they have no plan [...]' (*RTU Noticias*, 16 August 2011; www.youtube.com/watch?v=Cg2kpmFHrF8). This is ratified by an internal report by the SENAGUA (2012) stating that the environmental licensing and 'social participation' took place almost five months *after* signing the project construction contract.

Interestingly, the involved technicians were not at all blind to the rural communities' special ties to the river. A technician from the project's social and environmental area, for instance, acknowledged the territorial roots and incommensurable values of the water–society relationship by stating:

> I always say that this is a very complex issue to understand. Only someone who has lived on a river understands the river. We people from Chone are attached to our things, to our mountains, to our farms, to our rivers [...] you live on a river, where you bath, where you catch crayfish in the rainy season, where you water your horse, where you have been all your life, next to the river – and now they take the river away from you – that's very hard. These things have no price. (personal communication, 1 September 2015)

Yet, at the same time, he recognized that these factors were not taken into account in the decision-making process: 'what we didn't achieve during the social participation process was to agree on the dam construction by consensus, because there was nothing to agree upon – it was going to happen anyway'. This project justification went hand in hand with the subordination of rural interests, particularly in the zone affected by the dam.

The multipurpose project was mostly supported by Chone's urban population. A SENAGUA advisor acknowledged: 'the urban population supported the project wholeheartedly. In the rural zone, in Río Grande, they disagreed' (personal communication, 2 August 2015). Along the highway leading to the project zone were signs: 'Here we are building Chone's dream. Change is underway, and nothing can stop it.' With this backing, the construction contract was signed in July 2010 in the presence of national and local government authorities. At this event, inhabitants of Río Grande communities occupied the surroundings of the site and protested the project.

Río Grande communities are 'different' but 'potentially equal'

> Wanting to transform society through public policy and through efficiency and effectiveness is not possible, because true transformation is within human beings: we have to decode pre-modern patterns and encode new, modern patterns in order to construct Good Living. (ex-secretary of the National Water Secretariat, 21 July 2016)

As the former water secretary said, constructing development or *Buen Vivir* as a cross-cutting government goal was frequently based on an antagonistic duality: *pre-modern* versus *modern*. In practice, this is evident in portraying the city of Chone as a privileged space that must be protected from flooding, while the image of rural Río Grande region was (consciously or unconsciously) subordinated as the 'other', backward and even dangerous. As the SENAGUA technician put it, 'They are dangerous people. Especially my colleague in charge of expropriations has had to deal with people who have acted like criminals and threatened him.' Such characterization of rural reality contributed to legitimize the project. The latter would be replete of benefits for the dam-affected farmers, who would be enabled to 'improve'. That is, they could become developed 'just like folks in the city', and were therefore *potentially equal.*

They have already lived in isolation for all of their lives. Of course, you can't tell them this to their faces, because they would take it as an offense, because they would think that we technicians look down on them. [...] We have been generous enough to intervene. The project actually helped them improve their lives. They have always lived badly – you could say that they have been backward communities. (personal communication, SENAGUA's Socio-Environmental Department Manager, 1 September, 2015)

As mentioned, the dam implementation was not free of protests by Río Grande communities. Communities began organizing in 2009, based on an earlier organization devoted to providing order and local security and preventing the loss of livestock assets: the Central Committee of Rural Communities of Río Grande (CCRG). Most members were small and medium-scale farmers in the Río Grande watershed. This initial organization triggered the formation of the Defense Committee of Río Grande.

The resistance also included other stakeholders who were not CCRG members: critical scholars, non-governmental organizations (NGOs), local and national politicians, rural people affected by other dams, and critical citizens from the city of Chone. The resistance mobilized in several ways to make their complaints heard: judicial lawsuits, blocking the streets in the cities of Chone and Quito (the capital), symbolic demonstrations rejecting the dam, articles in the press, social networks and preventing outsiders from entering the Río Grande zone. All these activities postponed the beginning of the construction works and physically prevented the construction company from entering the zone. As the project moved forward, protests grew stronger, which made the dam construction begin over a year late. However, resistance was gradually demobilized by a range of government actions.

For example, as the opposition movement refused to abandon the site where they were gathering (the farm of one of the leaders, located precisely where the dam was ultimately built), the SENAGUA declared several farms as properties of public use. On that basis, a judge in Chone ordered the police to provide support and backing for immediate, violent occupation of the site where the resistance had gathered. The police report, which was part of an internal SENAGUA document, narrates:

> With the support of police personnel [...] we proceeded to enter that property for the immediate occupation ordered by the responsible authority. We were greater in number than the opponents, so we were able to dissuade them and get them to abandon the land, through progressive use of force. (SENAGUA, 2012, p. 267)

Within hours after the eviction, while the building company's machinery entered the zone to demolish homes and clear vegetation, in Quito the use of force was legitimized and the state's sovereign power was ratified by President Correa, by signing Decree No. 914, which declared the project zone an area of national security. Therefore, the police and army kept guard over the area to guarantee the project's implementation. The coercive strategy worked, according to SENAGUA (2012, p. 247): 'We must highlight the results achieved by the police entrance into the dam construction zone [...] which got the owners from the Río Grande zone who were opposing the dam to join in supporting it.'

Driven by the imperative to build the dam, the government used an antagonistic duality to justify implementing various coercive and non-coercive strategies to disperse the opposition. On the one hand, those against the dam were portrayed as radically opposed to progress. The then Vice-President even labelled them 'people-delayers' and 'political riffraff', meaning the opposite of development and Good Living. Or, as the former secretary of the SENAGUA said: 'It is a clash between change and traditions. [...] The rural folk didn't understand the future, and we wanted to show them the future. They are defending pre-modern ways of life' (personal communication, 21 July 2016). In that narrative, the opposition was defending almost irrationally different ways of life from the ways desired or prescribed by the government as *adequate*. On the other hand, several offers of compensation were tendered to the farmers in the resistance, based on the idea of urbanizing rural life (for a comparative 'commensuration/equalization' process, see Hoogendam & Boelens, 2019). In the government's view, rural communities needed to become modern and dignified, and the right path towards it was getting as close as possible to an urban lifestyle. An important part of this proposal included building the state's so-called 'model institutions': the 'millennium community' called Garden City (Ciudad Jardín) and a 'millennium school'. A SENAGUA advisor put it this way when referring to the resettlement programme: 'This was a great achievement. Building the multi-purpose project made it possible to propose a model city like the one that was built.'[2] That is, opponents who accepted the government's rules, values and ideology regarding what developed means, in this case urban, found the doors open to the opportunity to be a little more equal.

Governing society through water: Ecuador Estratégico (EP) and its urban–rural dream

The compensations implemented by the government took different shapes and had two purposes: first, to disperse the resistance, and second, to promote development for the people affected by the dam. The government-installed institute Ecuador Estratégico (EP) played a major role in both aims. It was created in 2011 as an institutional proposal to materialize the *Buen Vivir* discourse by investing in schools, millennium communities and paved roads in zones affected by major projects considered strategic.

In the conflict created by the dam in Río Grande, the millennium community and school were crucial parts of a radical intervention in rural communities: the project would help the backward (different) to become developed (equal). With this intention, the millennium community was planned and built to set up a civilized-urban lifestyle, targeting peasants on the properties affected. During the millennium community's planning stage, technicians approached farmers opposing the project on an individual basis to offer them houses on the condition that they persuade more people to desist from protesting. This strategy was effective: several farmers accepted the proposal. In the words of a former opposition leader and one of the first leaders of the millennium community:

> We were protesting [...] but one night an engineer from SENAGUA called me and said: 'you know, the President is going to build you a new little town. On such and such a date, bring all the people to sign, to give them these houses.' Then I went along, and I switched from the opposition to being in the government's favor. We gave in. We were blind, and others in the opposition, who now live in these new houses, were, too. (personal communication, 10 December, 2014)

Interestingly, during the early stages of planning compensatory measures, projectors considered creating three new resettlements right near the affected area on the shores of the lake reservoir (Figure 3). However, one aspect that largely guided final decisions about locating the resettlement village for affected families was the idea that the projected micro-watershed (where the dam, the reservoir and its shores were located) should remain an empty or empty-able place, mouldable according to the needs of optimal technical design without social interferences.[3]

Technicians saw the construction project as an opportunity to bring *order* to that previously disorganized territory. As one of the technicians responsible for the project's socio-environmental management put it:

> It was ideal to have the area empty of people, [...] we decided to locate the resettlement outside of the micro-watershed to avoid affecting the reservoir's water quality, because of waste or basically any environmental problems. President Correa wanted this to be a model resettlement for Latin America linked with the mega-project. (personal communication, 1 September, 2015)

Thus, the millennium community was built in the Mosquito River basin, outside the Río Grande micro-watershed (Figure 1). The multipurpose project was not only a way to control and organize nature but also a way to put society in order.

Once the community had been built, the EEEP established and publicized new rules and norms for living in this newly configured biophysical space. Ignoring local customs, the community was constructed as an example of how *Buen Vivir* is an urban lifestyle. In the middle of an entirely rural province, as an enclave, the community – like any urban neighbourhood – is organized into city blocks, with traffic signs and speed bumps every

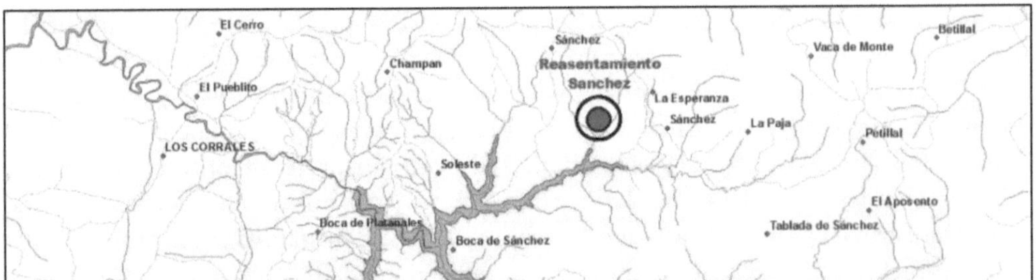

Figure 3. Original resettlement plan with three different locations around the reservoir area.

Source: The National Water Secretariat's (SENAGUA) environmental impact assessment (EIA) social participation material.

10 m. It also has community spaces: park, market, church, computing centre. The new rules for living together in the community include: organizing by blocks to maintain the common areas, not having parties later than midnight, not keeping poultry or hogs, and not leaving the house empty for longer than two weeks. If rules are not kept, the EEEP would analyze the possibility of giving the house to another family (Figures 4 and 5).

No house has any extra land to grow plants or raise animals; instead, the government purchased two farms to promote collective work by the resettled people. This, however, was based on a romanticized–socialist collectivization utopia and went against the everyday individual farm-work practices that these farmers had pursued for centuries (Ferrín-Schettini, 1986). The government envisioned the implementation of an urban life style for the inhabitants of Ciudad Jardín, simultaneously assigning the people the position of farmers, albeit under government-dictated conditions.

Despite the government's efforts, implementing coexistence rules was not enough to adapt these rural people to a more urban, rational lifestyle. As part of the process, the EEEP developed a board game to teach the resettled people the *right way* – according to the official political project – to achieve happiness, or *Buen Vivir*. Community promoters taught the community this game in several training sessions and visits to resettled people from other strategic projects (agricultural, inter-basin water transfer, dams, hydropower) constructed by the government. The game has three versions, one for each strategic sector: mining, hydrocarbons and hydropower megaprojects. The logic of the game is similar to the popular board game Monopoly. It includes drawings and information leaflets in simple, direct language, with dice and a path to follow, full of boxes that reward or punish on the basis of players' choices. As long as the player reaches boxes corresponding to the Citizens' Revolution government flagship projects, the pathway to happiness is guaranteed. A fundamental part of the game, as players advance on the board, is to answer a series of questions that openly promote the government's political project. Therefore, one leaflet asks: 'In Ecuador, do petroleum activities protect water resources?' The answer is 'Yes'. Another asks, 'What is the money from strategic resources used for?' The categorical answer is: 'To generate development for the country.' Perhaps one of the most suggestive questions is: 'What does petroleum mean to Ecuador?' The only answer is: 'Development, prosperity and well-being.' Through this game, the government's norms and truths are presented to community residents.

Figures 4 and 5. Ciudad Jardín millennium community. Photos: Authors.

A few months after moving to the community, especially women were worried about not having any work or any piece of land near their home, and they began raising chickens and a few hogs, disobeying the rules. Some women began setting up organic food gardens in their own homes; others began cultivating illegally on land outside the community limits. This gradual disobedience of the rules irked community leaders, who said: 'People don't follow the rules because the problem is that they are not used to living well' (personal communication, 5 November 2015). However, they had difficulties not only with living together in the community but also with working on the farms the government had purchased for them. Although the Ministry of Agriculture created an irrigators committee and trained farmers about the experience of community projects implemented in other regions of the country, this working modality was not welcomed and met with resistance (cf. Scott's, 1985, everyday resistance). A leader of the irrigators committee admitted about work on one of the farms:

> When the corn was fumigated, I saw that people did not collaborate […] they are lazy and don't understand. They didn't mix the chemical with the water in the pump, they just filled it with water halfway and pretended to spray, but they stole the liquid or the fertilizer, sneaking it out in the disposable containers that lunch came in. (personal communication, 5 November, 2015)

This behaviour could be explained partly because many resettled people still had working relationships with their previous employers or neighbours. Some still had some land where they used to live. Many preferred to ride a mule for over an hour to the expropriated land (not flooded yet) to continue working there, instead of participating in collective work on the new farms.

Through all these changes, the norms and discourses for the dam have reconfigured not only the landscape and its ecology but also the social, agro-productive and working relationships of local communities. The governor of the province of Manabí, in her inaugural speech, encapsulated the meaning of this mega-project:

> This project has a soul and a body. The body is this majestic dam, but the soul is being planted in the consciousness, in the soul of the purest people: in the souls of our girls and boys. This project is not only infrastructure. (24 November, 2015)

However, over 1000 ha were permanently flooded in the Río Grande zone, sacrificing rural communities and families' homesteads. Beyond the environmental impact, nearly 100 rural families have not been resettled, their lands, livelihoods and histories have been flooded by the water governance/dam design decisions – many being left in deeply precarious conditions. Several face difficulties at reframing their livelihoods and accessing their territory because of the water lilies building up on the reservoir's surface, which obstructs their ways of working in different territorial sites and significantly increases transport costs. Further, today there is an evident social division even among those who once were united and belong to the resistance. The mega-project was presented as a technical solution to diminish floods risk, while its soul promoted, legitimized and justified an official sociopolitical project contrasting rural and urban life,[4] and governance of the abundance of water.

Discussion and conclusions

We have shown in this paper how the discourse on water abundance is used to promote mega-dams as technical, apolitical remedies, which, however, heighten relationships of inequity between urban and rural zones. We have also shown how the solution chosen in Chone follows a national policy of believing in modern living as a stepping stone toward development, or *Buen Vivir*. The ways in which the Citizen's Revolution government uses the concept of *Buen Vivir* is highly controversial in this case. Regardless of its origins, the government has appropriated the concept, endowing it with the government's own notions of development and underdevelopment (Van Teijlingen & Hogenboom, 2016). It has been used as to legitimize the government's political water projects and its 'postcolonial conditions of development' (Radcliffe, 2012, p. 240). Such modernizing policy about life and living, and its programmes and projects in the water sector, have been backed by the comeback of a wolf state in a sheep's clothing. Its reforms, proclaimed under the banners of participation and inclusion, have been thoroughly infused with a hierarchical ranking of knowledge (modern versus pre-modern), forms of governance (good versus traditional), ideologies (Socialism for Good Living), territories (development versus sacrifice), persons (majorities versus sacrificed minorities) and geographical spaces (urban versus rural). Under these circumstances, the city of Chone bolstered its image as a development territory, while Río Grande and its rural communities were labelled backward and sacrifice-able. This has meant that, by building the dam, the city is now partially protected from flooding, because the Río Grande's water has been controlled. However, much of the city and surrounding areas continue to flood when other tributaries, such as the Garrapata, overflow. Soon, this 'technical deficiency' may be used to justify the construction of other mega-infrastructures, as happens in other Ecuadorian hydrosocial territories.[5] Meeting initial objectives creates the need to plan and implement more projects following on from the first one, justified by the promise of greater efficiency.

Mega-dams such as Chone's are commonly justified under the utilitarian premise: *the well-being of the majority*, but it is the (rural) majorities who actually are most affected by these policies. The social and political construction of 'majorities' and 'minorities' follows the ways in which higher political–economic value is given to urban zones or elite properties. These areas have priority to be protected from flooding risks – despite the fundamental socioeconomic importance of the rural areas for *most* people's livelihoods. According to the distribution of inhabitants among rural and urban areas, the 'majorities' discourse that the government used as a justification to build the dam was not totally accurate. This case shows that the majority actually live in rural areas. Therefore, the utilitarian myth obeys a hierarchical ranking to suit the interests of those in power. As we have shown, this becomes manifested in how the notion of water overabundance is defined and informed by economic and political powers rather than just by biophysical and natural conditions or any objective technical considerations.

We have also analyzed how this hierarchical construction not only justifies construction of megaprojects such as Chone's but also – and structurally – (re)produces profound damage to and transformations of rural societies. We go beyond an analysis of majorities versus minorities. It reveals a more structural aspect of utilitarianism, which is labelling people, territories, ideologies and ways of life as radically different, but also potentially equal whenever they obey power and behave according to as modern defined norms and ways of life. This, as we have shown in Chone, subjects the so-called minorities to transformation. Consequently, by following the norms, values, ideologies and lifestyles pre-designated by those in power, those minorities – for instance, in millennium communities – can become *just as equal.* That is, if they would follow the right rules, rural people can be just as developed, modern and happy as urban people. As an affected woman expressed: 'We are just happy here, now we live like in a gated community of Guayaquil. Why shouldn't I be happy?'

The case of water overabundance contributes to an urban–rural political ecology conceptualization. It emphasizes that the linkage between rural and urban is not only aggravated by spreading the city limits out into the countryside, absorbing rural space, resources and inhabitants. Complementarily, this encroachment is fed by cultural politics, the clashes between different, opposing categorizations of superiority and inferiority, historically granted to urban and rural settings.

Understanding the social construction and legitimizing discourses of the black boxes *water scarcity* and *water abundance* opens our eyes to scrutinize issues of inequity in water governance. In our case, the water-abundance discourse, and flood control for cities, encloses and reproduces particular forms of water injustices. It especially shows, by examining permanently drowned waterscapes, how rural waters are not simply materially transformed but also socially reordered and controlled.

Acknowledgements

We thank Justicia Hídrica collegues for the feedback provided on the first drafts of this paper. We also thank the editors and anonymous reviewers for their insightful and useful comments. We give special recognition to the Río Grande social movement and to those that let us stay in their houses during the field work phase.

Notes

1. Ecuador's political–administrative division has four nested subnational levels: province, canton, parish and neighbourhoods (urban) or hamlets and communities (rural).
2. On the governmentality politics of these modern model villages ('Millennium Communities and Schools') in Ecuador, see Valladares and Boelens (2017).
3. According to SENAGUA, 181 families were directly affected by the dam construction; however, the millennium community has capacity for only 81. That is, nearly 55% of affected families have not been relocated.
4. See Özok-Gündoğan (2005) for a similar strategy implemented in Turkey.
5. We show such dynamic in the case of the Daule-Peripa Dam, which was constructed in the early 1980s in coastal Ecuador (Hidalgo-Bastidas et al., 2018).

Disclosure statement

No potential conflict of interest was reported by the authors.

Funding

This work was supported by the Secretaría Nacional de Educación Superior, Ciencia, Tecnología e Innovación (Ecuador).

ORCID

Juan Pablo Hidalgo-Bastidas http://orcid.org/0000-0002-0068-4671
Rutgerd Boelens http://orcid.org/0000-0002-8412-3109

References

Acosta, A., Álvarez-Gonzáles, F., Ávila-Santamaría, R., Castro-Rieria, C., Cuvi, J., Dávalos, P., De la Torre, C., ... Villavicencio, A. (2013). *El correísmo al desnudo* (*Segunda* ed.). Quito: Montecristi Vive.

Aguilera-Klink, F., Pérez-Moriana, E., & Sánchez-García, J. (2000). The social construction of scarcity. The case of water in Tenerife (Canary Islands). *Ecological Economics*, *34*(2), 233–245. doi:10.1016/S0921-8009(00)00160-9

Alatout, S. (2009). Bringing abundance into environmental politics: Constructing a Zionist network of water abundance, immigration, and colonization. *Social Studies of Science*, *39*(3), 363–394. doi:10.1177/0306312708101979

Baghel, R. (2014). *River control in India: Spatial, governmental and subjective dimensions.* Dordrecht: Springer Science & Business Media.

Bakker, K. (2010). *Privatizing water: Governance failure and the world's urban water crisis.* New York: Cornell University Press.

Banister, J. M., & Widdifield, S. G. (2014). The debut of 'modern water' in early 20th century Mexico City: The Xochimilco potable waterworks. *Journal of Historical Geography*, *46*, 36–52. doi:10.1016/j.jhg.2014.09.005

Baviskar, A. (2007). *Waterscapes. The cultural politics of a natural resource.* Delhi: Permanent Black.

Boelens, R. (2015a). *Water justice in Latin America. The politics of difference, equality, and indifference.* Amsterdam: Center for Latin American Research and Documentation (CEDLA)/University of Amsterdam.

Boelens, R. (2015b). *Water, power and identity: The Cultural politics of water in the Andes.* Washington, DC, and London: Earthscan and Routledge.

Boelens, R., Hoogesteger, J., & Baud, M. (2015). Water reform governmentality in Ecuador: Neoliberalism, centralization and the restraining of polycentric authority and community rule-making. *Geoforum*, *64*, 281–291. doi:10.1016/j.geoforum.2013.07.005

Duarte-Abadía, B., & Boelens, R. (2016). Disputes over territorial boundaries and diverging valuation languages: The Santurban hydrosocial highlands territory in Colombia. *Water International*, *41*(1), 15–36. doi:10.1080/02508060.2016.1117271

Duarte-Abadía, B., & Boelens, R. (2019). Colonizing rural waters. The politics of hydro-territorial transformation in the Guadalhorce Valley, Málaga, Spain. Special Issue 'Rural–urban water struggles'. *Water International*, *44*(2), 148–168. doi:10.1080/02508060.2019.1578080

Erensu, S. (2013). Abundance and scarcity amidst the crisis of 'modern water': The changing water–energy nexus in Turkey. In L. Harris, J. Goldin, & C. Sneddon (Eds.), *Contemporary water governance in the Global South: Scarcity, marketization and participation* (pp. 61–78). New York: Routledge.

Ferrín-Schettini, R. (1986). *Economías campesinas, estructura agraria y formas de acumulación: El caso de Manabí a partir de la Revolución Liberal.* Quito: PUCE-CONUEP.

Goldman, M. (2007). How 'Water for All!' policy became hegemonic: The power of the World Bank and its transnational policy networks. *Geoforum*, *38*, 786–800. doi:10.1016/j.geoforum.2005.10.008

Gudynas, E., & Acosta, A. (2010). Si eres tan progresista? Por qué destruyes la naturaleza? Neoextractivismo, izquierda y alternativas. *Ecuador Debate*, *79*(5), 61–82.

Hidalgo, J. P., Boelens, R., & Vos, J. (2017). De-colonizing water. Dispossession, water insecurity, and Indigenous claims for resources, authority, and territory. *Water History*, *9*, 67–85. doi:10.1007/s12685-016-0186-6

Hidalgo-Bastidas, J. P., & Boelens, R. (2019). Hydraulic order and the politics of the governed. The Baba Dam in coastal Ecuador. *Water*, *11*(3), 409; doi:10.3390/w11030409

Hidalgo-Bastidas, J. P., Boelens, R., & Isch, E. (2018). Hydro-territorial configuration and confrontation. The Daule-Peripa multi-purpose hydraulic scheme in coastal Ecuador. *Latin American Research Review*, *53*(3), 517–534. doi:10.25222/larr.362

Hommes, L., & Boelens, R. (2017). Urbanizing rural waters: Rural–urban water transfers and the reconfiguration of hydrosocial territories in Lima. *Political Geography*, *57*, 71–80. doi:10.1016/j.polgeo.2016.12.002

Hommes, L., Boelens, R., Duarte-Abadía, B., Hidalgo-Bastidas, J. P., & Hoogesteger, J. (2018). Reconfiguration of hydrosocial territories and struggles for water justice. In R. Boelens, T. Perreault, & J. Vos (Eds.), *Water justice* (pp. 151–168). Cambridge, UK: Cambridge University Press. doi:10.1017/9781316831847.010

Hommes, L., Boelens, R., Harris, L., & Veldwisch, G. J. (2019). Rural–urban water struggles: Urbanizing hydrosocial territories and evolving connections, discourses and identities. *Water International*, *44*(2), 81–94. doi:10.1080/02508060.2019.1583311

Hoogendam, P., & Boelens, R. (2019). Dams and Damages. Conflicting epistemological frameworks and interests concerning "compensation" for the Misicuni project's socio-environmental impacts in Cochabamba, Bolivia. *Water*, *11*(3), 408. doi:10.3390/w11030408

Hoogesteger, J., Boelens, R., & Baud, M. (2016). Territorial pluralism: Water users' multi-scalar struggles against state ordering in Ecuador's highlands. *Water International*, *41*(1), 91–106. doi:10.1080/02508060.2016.1130910

Ioris, A. (2016). Water scarcity and the exclusionary city: The struggle for water justice in Lima, Peru. *Water International*, *41*(1), 125–139. doi:10.1080/02508060.2016.1124515

Johnston, B. R. (2018). Large-scale dam development and counter movements: Water justice struggles around Guatemala's Chixoy Dam. In R. Boelens, T. Perreault, & J. Vos (Eds.), *Water justice* (pp. 169–186). Cambridge: Cambridge University Press.

Linton, J., & Budds, J. (2014). The hydrosocial cycle: Defining and mobilizing a relational–dialectical approach to water. *Geoforum*, *57*, 170–180. doi:10.1016/j.geoforum.2013.10.008

Lynch, B. (2013). River of contention: Scarcity discourse and water competition in Highland Peru. *Georgia Journal of International and Comparative Law*, *42*, 115.

Lynch, B. (2019). What Hirschman's hiding hand hid in San Lorenzo and Chixoy. *Water*, *11*(3), 415. doi:10.3390/w11030415

Marks, D. (2019). Assembling the 2011 Thailand floods: Protecting farmers and inundating high-value industrial estates in a fragmented hydro-social territory. *Political Geography*, *68*, 66–76. doi:10.1016/j.polgeo.2018.10.002

McCully, P. (2001). *Silenced rivers: The ecology and politics of large dams*. Michigan: Zed Books.

Mehta, L. (2006). Whose scarcity? Whose property? The case of water in western India. *Land Use Policy*, *24*(4), 654–663.

Mena-Vásconez, P., Boelens, R., & Vos, J. (2016). Food or flowers? Contested transformations of community food security and water use priorities under new legal and market regimes in Ecuador's highlands. *Journal of Rural Studies*, *44*, 227–238. doi:10.1016/j.jrurstud.2016.02.011

Molle, F., & Berkoff, J. (2006). *Cities versus agriculture: Revisiting intersectoral water transfers, potential gains and conflicts. Comprehensive Assessment Research Report 10*. Colombo: IWMI.

Nixon, R. (2010). Unimagined communities: Developmental refugees, megadams and monumental modernity. *New Formations*, *69*(1), 62–80. doi:10.3898/NEWF.69.03.2010

Osti, G. (2017). The anti-flood detention basin projects in northern Italy. New wine in old bottles? *Water Alternatives*, *10*(2), 265–282.

Özok-Gündoğan, N. (2005). *'Social development' as a governmental strategy in the Southeastern Anatolia* project. *New Perspectives on Turkey*, *32*, 93–111.

Porter, J., & Demeritt, D. (2012). *Flood-risk management*, mapping, and planning: The institutional politics of decision support in England. *Environment and Planning*, *44*(10), 2359–2378. doi:10.1068/a44660

Radcliffe, S. (2012). Development for a postneoliberal era? Sumak Kawsay, living well and limits to decolonization in Ecuador. *Geoforum*, *43*, 240–249. doi:10.1016/j.geoforum.2011.09.003

Radcliffe, S., & Westwood, S. (2005). *Remaking the nation: Identity and politics in Latin America*. London and New York: Taylor & Francis.

Riaz, K. (2002). Tackling the issue of rural–urban water transfers in the Ta'iz region, Yemen. *Natural Resources Forum*, *26*(2002), 89–100. doi:10.1111/1477-8947.00010

Roa-García, M. C. (2014). Equity, efficiency and sustainability in water allocation in the Andes: Trade-offs in a full world. *Water Alternatives*, *7*(2), 298–319.

Romano, S. T. (2017). Building capacities for sustainable water governance at the grassroots: 'Organic empowerment' and its policy implications in Nicaragua. *Society & Natural Resources*, *30*(4), 471–487. doi:10.1080/08941920.2016.1273413

Roth, D., Khan, M. S. A., Jahan, I., Rahman, R., Narain, V., Singh, A. K., [...] Yakami, S. (2018). Climates of urbanization: Local experiences of water security, conflict and cooperation in peri-urban South Asia. *Climate Policy*, 1–16. doi:10.1080/14693062.2018.1530967

Roth, D., & Warner, J. (2007). Flood risk, uncertainty and changing river protection policy in the Netherlands: The case of 'calamity polders'. *Tijdschrift voor economische en sociale geografie*, *98*(4), 519–525.

Scott, J. C. (1985). *Weapons of the weak : Everyday forms of peasant resistance*. New Haven London: Yale University Press.

SENAGUA. (2012). *Informe detallado de los puntos solicitados por la procuraduría general del estado para la defensa del Estado ecuatoriano ante la CIDH*. Quito: SENAGUA.

Shah, E., Liebrand, J. W., Vos, J., Veldwisch, G. J., & Boelens, R. (2018). The UN water development report 2016 'Water and Jobs': A critical review. *Development and Change*, *49* (2), 678–691. doi:10.1111/dech.12395

Swyngedouw, E. (1995). The contradictions of urban water provision: A study of Guayaquil, Ecuador. *Third World Planning Review*, *17*, 4.

Swyngedouw, E. (2004). *Social power and the urbanization of water: Flows of power*. Oxford: Oxford University Press.

Torio, P., Harris, L., & Angeles, L. (2019). The rural–urban equity nexus of Metro Manila's water system. *Water International*, *44*(2), 115–128. doi:10.1080/02508060.2019.1560559

Valladares, C., & Boelens, R. (2017). Extractivism and the rights of nature: Governmentality, 'convenient communities' and epistemic pacts in Ecuador. *Environmental Politics*, *26*(6), 1015–1034. doi:10.1080/09644016.2017.1338384

Van Teijlingen, K. (2016). The 'will to improve' at the mining frontier: Neo-extractivism, development and governmentality in the Ecuadorian Amazon. *The Extractive Industries and Society*, *3*(4), 902–911. doi:10.1016/j.exis.2016.10.009

Van Teijlingen, K., & Hogenboom, B. (2016). Debating alternative development at the mining frontier: Buen Vivir and the conflict around El Mirador mine in Ecuador. *Journal of Developing Societies*, *32*(4), 382–420. doi:10.1177/0169796X16667190

Warner, J. (2010). *Flood planning: The politics of water security*. London: I. B. Tauris.

Warner, J., Hoogesteger, J., & Hidalgo-Bastidas, J. P. (2017). Old wine in new bottles: The adaptive capacity of the hydraulic mission in Ecuador. *Water Alternatives*, *10*(2), 322–340.

OPEN ACCESS

Upsetting the apple cart? Export fruit production, water pollution and social unrest in the Elgin Valley, South Africa

Matthijs Wessels, Gert Jan Veldwisch, Katarzyna Kujawa and Brian Delcarme

ABSTRACT

This article explores the encounter between two contrasting visions of how the hydrosocial territory of the Elgin Valley of South Africa is, and should be, constituted and the conflicts over water pollution this gives rise to. It studies how poor urban dwellers try to upset the status quo of unequal access to land and water, which is linked to broader, historically entrenched, inequalities. White commercial farmers have succeeded in upholding the dominant hydro-territorial order by emphasizing the economic importance of their sector, by reducing complex political issues to technical challenges, and by capturing 'democratic' water institutions.

Introduction

In this article we study the interactions between white commercial fruit farmers and poor (coloured and black) urban dwellers, which we present as a contested process of reconfiguring hydrosocial territory – a confrontation between two contrasting visions of the Elgin Valley and its water use practices, but also about the ecological, socio-economic, institutional-political and symbolic spaces with which they are intertwined. To that end we study how the status quo of unequal access to resources is protected and confronted. In particular, we enquire how urban water pollution and social unrest contribute to a push to 'upset the apple cart', i.e., disturb the status quo, and how farmers defend their vested interests.

The Elgin Valley, in the Western Cape, South Africa, is characterized by strong differences in access to natural resources and public services that reflect the country's apartheid history. The fertile lands of the Elgin Valley have long been occupied by economically well-off white South Africans, who grow deciduous fruit (mainly apples and pears) for the international market. In contrast, the urban poor, living in the informal areas of the town of Grabouw, are largely black and coloured families living with limited or no economic opportunities. Fruit production in the valley, in particular during the dry summer period, depends largely on irrigation and water flows in the Palmiet River, which are controlled through six in-stream storage dams. With urban expansion, wastewater discharges have degraded the water quality in the Palmiet, and pathogens (*Escherichia coli*, *E. coli*) in the water used for irrigation pose a threat to farmers' access to export markets, given the strict water quality requirements of their customers.

While studying the issue of water pollution and the reuse of wastewater for irrigation we realized that not only the pollution itself was contested, but also the meaning of the pollution: defining pollution problems, sources and responsibilities, as well as corresponding solutions, is a deeply political process, though often presented as merely a practical problem. Along these lines, we observe a process whereby complex social and political processes in society are 'rendered technical' to simplify them and make it legitimate to pursue 'business as usual' (Ferguson, 1994; Li, 2007). In the process, alternatives to the existing socio-natural orderings are sidelined. Complex situations, co-constituted by social, natural and political processes, are reduced to mere biophysical problems that can be solved 'objectively', concealing the often deeply political nature of the proposed solutions (Boelens, Hoogesteger, Swyngedouw, Vos, & Wester, 2016).

The case of water quality issues provides an excellent lens to study rural–urban struggles and can be understood as a process of contesting and seeking to reconfigure hydrosocial territories. Hydrosocial territories are 'socially, naturally and politically constituted spaces that are (re)created through the interactions among human practices, water flows, hydraulic technologies, biophysical elements, socio-economic structures and cultural-political institutions' (Boelens et al., 2016, p. 1; see also Hommes, Boelens, & Maat, 2016; Hoogesteger, Boelens, & Baud, 2016; Swyngedouw & Boelens, 2018; Swyngedouw & Williams, 2016). In (re)creating these configurations, also called 'hydrosocial networks' or 'waterscapes' (Linton & Budds, 2014; Swyngedouw, 1999), actors compete to establish a hydro-territorial order that is in line with their interests and convictions. As interests and territorial projects are almost always diverse and at times contradict each other, the presence of different actors and their socio-territorial projects may lead to 'territorial pluralism', in which diverse territories overlap and interact in one and the same geographical-political space (Hommes, Boelens, Harris, & Veldwisch, 2019; Hoogesteger et al., 2016).

The fieldwork for this article was undertaken between November 2015 and February 2016, with the primary aim of understanding the effects of urban water pollution on irrigation in the Elgin Valley (Wessels, 2016). Water quality analyses, in-depth stakeholder interviews, and field visits were combined in this analytical case study research. The water quality analyses were drawn from three monitoring programmes in the Elgin Valley that take year-round (mostly monthly) repeat samples at specific locations throughout the valley. The results of these analyses are only used where relevant to the focus of this article. Semi-structured interviews were used to learn about the linkages between urban water pollution and irrigation practices. We interviewed 12 commercial farmers who depend on water from the Palmiet River. Other important stakeholders who were interviewed and visited on a regular basis included the Theewaterskloof Municipality, the Groenland Water User Association (GWUA) and the Breede-Gouritz Catchment Management Agency (BGCMA). Triangulation of sources was used wherever possible to verify and validate water quality data and the interview results. Field visits to farms, sewerage and treatment infrastructure, and urban settlements were also undertaken, the latter specifically to monitor possible sewerage spillages. After completion of the field work, the research focus shifted to also incorporate the political aspects of water management and urban–rural interactions. The field research results have therefore been complemented by desk research on the politics of (urban) water governance in South Africa, with specific reference to the Elgin Valley. We realize the limitations of using secondary sources for a part of our analysis and throughout the text indicate the sources on which our analysis is based.

The present-day urban–rural water struggles in the Elgin Valley are inseparably linked to the broader political history of access to land, water and labour in South Africa. Therefore, the following section presents an analysis of the history and current characteristics of the fruit sector in the Elgin Valley as the establishment of a hydrosocial territory. Thereafter, we explore an alternative hydrosocial perspective: that of the urban poor in Grabouw. We then analyze how urban water pollution threatens the dominant order and how farmers have guaranteed its continuity, despite water reforms that aimed to undermine their power. In the final section we draw conclusions about the reconfiguration of the valley's hydrosocial territories along the rural–urban axis, showing how these remain inextricably tied to historical patterns of power, wealth and race.

The establishment of hydrosocial territory

The Elgin Valley is one of the most important deciduous fruit producing and exporting areas of South Africa, especially for apple production (Hortgro, 2016; TWK, 2015c). In this section we first analyze this sector, its history and main actors, and then show how (predominantly) white farmers perceive and portray the valley's productivity in relation to its land and water economy, crucially positioning commercial fruit farming at the centre of a hydrosocial territory with highly skewed burdens and benefits.

Access to land plays an important role in shaping patterns of income and employment in rural areas, and largely determines the distribution of wealth, status and authority in society (Cohen, 1980, cited in Mabaya, Tihanyi, & Karaan, 2011). Considering that land access, in South Africa and elsewhere, is deeply entwined with the country's political history, it is indispensable to understand the present-day agricultural sector within its historic context. In South Africa, land acquisition by white settlers during the twentieth century happened under a government focused on consolidating power in the hands of the white minority. This enabled the Afrikaner (white) farmers to secure land ownership and develop large-scale, commercial and mechanized farms (Mabaya et al., 2011; Vink & Kirsten, 2003). Black South Africans lost their land rights and became redundant to the labour needs of the commercial and mechanized farms (Thompson, 1990). In the post-apartheid period of the past several decades, land reform processes were initiated 'to return land to people who were forcibly evicted in the past and compensate them; redistribute land to people who suffered discrimination; and improve land tenure security for farm workers and others' (Boudreaux, 2010, p. 15) to create an environment conducive for them to establish emerging farms. Reform processes, however, have been slow and fallen short of both official targets and public expectations (Boudreaux, 2010; Hall, 2004; Lahiff & Cousins, 2005; Mabaya et al., 2011). Voluntary redistribution of agricultural land, as part of the Black Economic Empowerment programme, in South African commercial agriculture has had little impact to structurally transform the conditions of the people disadvantaged by unfair discrimination, and historically disadvantaged individuals (HDIs) remain enmeshed in the inheritance of inequality (Bernstein, 2013; Du Toit, Kruger, & Ponte, 2008).

In terms of water use, in 2011 the Catchment Management Agency (CMA) reported that the portion of water used by HDIs is between 2% and 5% of the water available in the Breede-Overberg Water Management Area. Plans are to increase the agricultural land owned by HDIs to 15% within five years and ultimately to 30% over a period of 15 years. These land reform processes also require the re-allocation of water rights to emerging farmers (BGCMA, 2017; BOCMA, 2011). It is unclear what the current status is, but white commercial farmers in the Elgin Valley continue to hold a strong position with regard to access to land, water, labour, knowledge and financial means, essential elements of the dominant hydro-territorial order. This organization of agriculture in relations of production in the Elgin Valley reflects the history of land tenure and commercialization, so that today the farms in the valley are predominantly Afrikaner owned and managed.

As irrigation is necessary for agricultural production during the hot and dry summer, since the 1970s several dams have been built along the 70 km Palmiet River and its 11 tributaries to secure year-round water availability for irrigation of large-scale farms (Figure 1). Farmers perceive the construction of the dam and irrigation infrastructure as a historic driver of change in the region (Peters & Wilson, 2007). The largest dam (Eikenhof Dam), located upstream of Grabouw, supplies water to most of the fruit farmers (just over 75% of the total irrigated area in the catchment) via an underground piped system. Other farms, especially those downstream of Grabouw, are not connected to the piped system and draw water directly from the Palmiet River (Bosch, 2008; DWAF, 2004). These farmers have relatively little control over the quantity and especially the quality of the water they abstract. The agricultural community is by far the largest water user in the valley, cultivating 66% of the total catchment area and using 69% of the water (BOCMA, 2011). The GWUA is the main authority responsible for the day-to-day water supply and is mandated, per the South African National Water Act (Act 36 of 1998), to charge levies for water use. Water is allocated to different users, among which are commercial farmers, municipalities, industries and a small group of emerging farmers from historically disadvantaged communities (Baleta, 2015; GWUA, 2014). Water from the Eikenhof Dam is also used for domestic and industrial purposes in Grabouw and is transferred out of the valley to the nearby city of Cape Town. According to the CMA (BOCMA, 2011), around 30% of the water in the Palmiet Catchment is used by the Western Cape Water Supply System, which serves the larger Cape Town metropolitan area. Increasing demands from Cape Town and extended dry spells over the past years have put more pressure on the availability of water in the area (BGCMA, 2017).

As mentioned above, the commercial fruit farms in the Elgin Valley are primarily Afrikaner-managed farms, owned by local families or larger companies that have several farms in the Western Cape Province and beyond. In their hydro-political mapping of the valley and its actors, Peters and Wilson (2007) found that farmers present themselves as having lived in the valley for a long time. In our interviews, too, they referred to how the histories of their farms and families, some tracing back for several generations, are connected to the land they cultivate and the infrastructural and economic development of the valley, an important basis for their hydro-territorial claims. The farm managers we interviewed often have an academic degree in agriculture and make use of external consultants for technical issues such as the management of nutrients, pesticides and water. Their orchards and vineyards are generally equipped

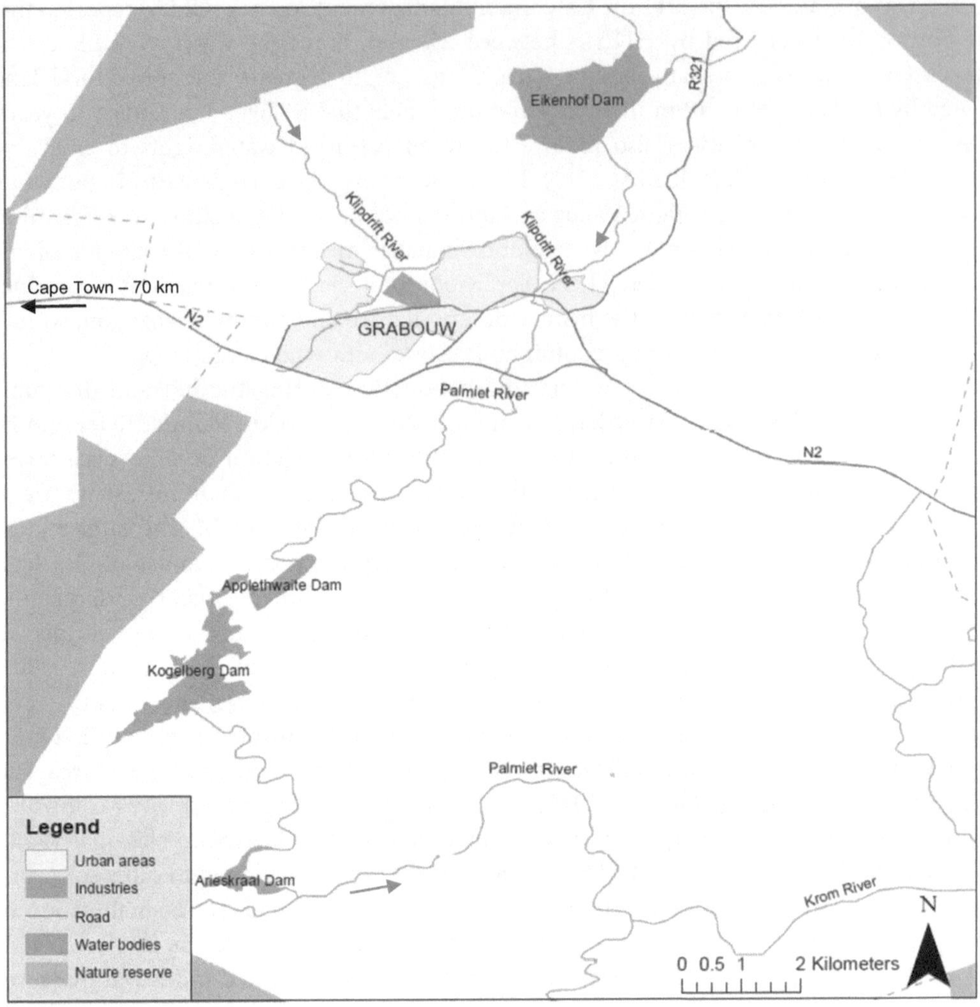

Figure 1. Elgin valley study area.
Source: own elaboration

with automated drip irrigation or micro-sprinkler systems and innovations such as new cultivars and state-of-the-art fruit-picking platforms that increase yields while reducing costs and labour dependency during the fruit-picking season.

The harvesting season runs from early January to early May, depending on the apple or pear variety and whether the season is early or late. The harvesting season is particularly labour-intensive because the time window for picking high-quality fruits is short. The estates, which range in size from 20 to 450 hectares, have yields of 60–100 t/ha of apples per year, largely sold on the global market. The harvesting of high-quality fruits during this short period requires large and reliable labour inputs. Farmers contract cheap seasonal labourers (often black African workers) from the Eastern Cape Province, who are accommodated in hostels on the farm and organized in picking teams. After the picking season, the companies arrange transport for the workers to return home. However, in practice, many of the seasonal workers remain in the valley to seek other jobs in the fruit industry. In the first post-apartheid decade the sector saw a strong tendency towards the 'casualisation, externalisation and contractualisation of labour' (Ewert & Du Toit, 2005, p. 327), which effectively meant an increase in the number and share of casual labourers, re-employment under poorer conditions and the use of intermediaries for recruiting labour. Ewert and Du Toit (2005) interpret this tendency as a response to the insecurities that white farmers experienced as a result of post-apartheid governmental policies. In our interviews farmers said that they prefer to bring in new workers from outside the region every year as they are afraid that labourers from the urban settlements are more likely to initiate collective strikes during the picking season, narrowing or closing the farmers' window for picking high-quality fruits. Using contracted seasonal workers gives the farmers a stronger position over labour organization and costs (Baleta, 2015; Haysom, 2007). According to Haysom (2007, p. 25), the seasonal workers 'are often the most disadvantaged, earning arguably the lowest salaries in South Africa, and having the least access to benefits'.

Most farmers are members of a growers' cooperative, through which they share a packing facility and do joint branding and marketing. In the past, the United Kingdom and mainland Europe were the primary markets for top-grade fruit, but the markets have been changing over the past decade, partly due to the economic crisis in Europe. Farmers who produce for export markets must follow strict guidelines on fertilizer and pesticide use, labour ethics and water quality, which are set by GlobalGAP (2016). These standards are more stringent than the domestic requirements and are required for export and the farmers' economic gains. This makes water pollution a sensitive topic that local farmers are reluctant to talk about. Several water users interviewed referred to the uproar caused by a 2014 article, 'E coli in Grabouw water', on a South African news platform (News24, 2014), which claimed that 'high levels of pollution, including E. coli, have been found in the water in Grabouw in the Western Cape'. The article was based on a report from the CMA and was commented on by one of their officials. The article traced the pollution back to the town of Grabouw and argued that immediate action was needed 'to take precautionary measures and prevent damage to vital exports'. The GWUA argued that this was based on misinterpreted data, but irrespective of the strength of the empirical foundation, the article had already raised concerns among international fruit distributors. Fortunately, the concerns could be refuted, and access to Western supermarkets remained secure. The uproar caused by the article reflects the vulnerability of the agricultural sector with regard to water pollution and the potential threat to the commercial farmers' interests if they should lose control over water management and its representation.

Commercial fruit farmers stress the strategic importance of the fruit sector for the valley, both historically and for its future. In that context, access to sufficiently clean water is portrayed as a crucial element for the valley's vibrant economy, on which large numbers of people depend, directly or indirectly. This forms the basis of the dominant hydrosocial territory. As we explore further in the section below, however, this ordering involves a highly skewed distribution of burdens and benefits. Considering diverse perspectives, particularly those of local coloured and black town dwellers, adds different perspectives and understandings to these issues, and presents alternative hydrosocial orderings worthy of consideration.

An alternative hydrosocial perspective: the urban poor

Not everybody benefits equally from this economy, as can be observed in Grabouw, the valley's main urban centre. In this section we consider, on the basis of an analysis of secondary sources, how urban populations perceive and portray the fruit sector and its role in the hydrosocial ordering of the valley. It is shown that urban residents emphasize sanitation services and economic opportunities and thereby present a different hydrosocial perspective. Issues of water pollution from urban areas are regarded as a result of the shortcomings in sanitation services, and protests are used to challenge the inequalities in the valley.

With a population of over 30,000, Grabouw is the largest town in the valley (belonging to the Theewaterskloof Municipality) and borders the city of Cape Town (Figure 2), with a community of predominantly coloured (56%) and black people (39%).[1] In many ways it is similar to other agricultural areas in the Western Cape. Despite economic reforms to redress historic inequalities, income distribution remains strongly correlated with race (Haysom, 2007; Leibbrandt & Levinsohn, 2011). Unemployment in the Theewaterskloof Municipality is estimated at 14.9% (StatsSA, 2011), and 13.7% of the households earn less than R400 per month (about USD 1 per day) (TWK, 2016). These figures, however, are subject to the in-migration of labourers and the seasonality of the employment in the valley. In the off-season, unemployment

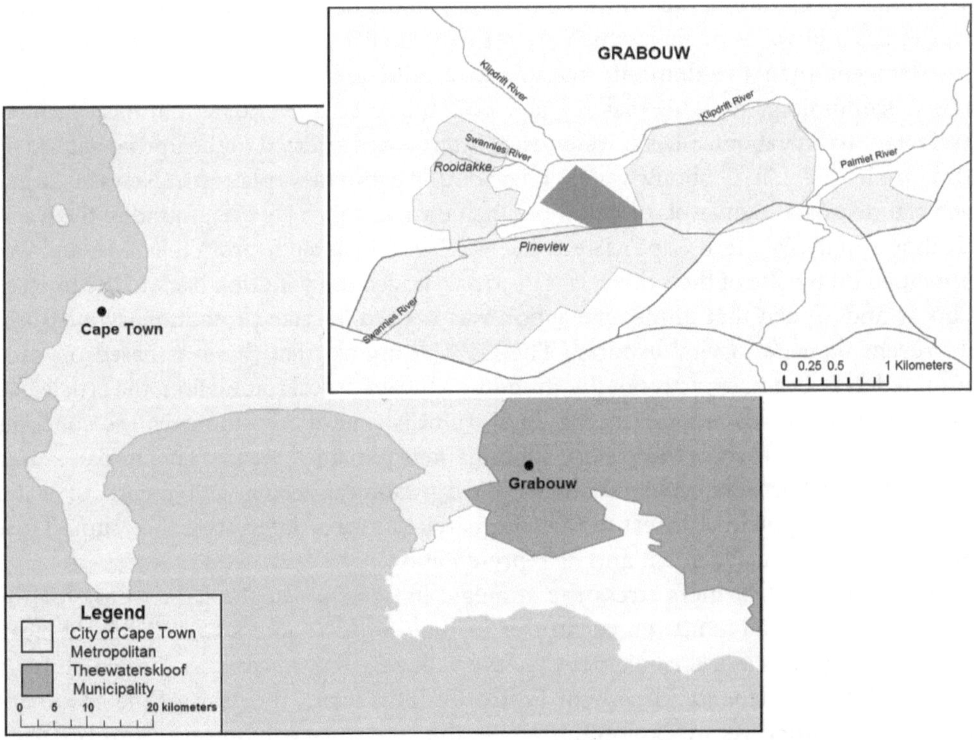

Figure 2. Location and map of Grabouw in the Theewaterskloof Municipality, bordering to the City of Cape Town Municipality.

Source: own elaboration

rates are much higher, and households in the informal areas have hardly enough food or money to live on (Haysom, 2007; Peters & Oldfield, 2005). There is a high dependence on the labour opportunities directly or indirectly linked to the fruit sector. However, the tendency of the commercial farmers to bring in new seasonal workers every year results in growing unemployment in the low-income areas and limited economic opportunities for residents. The municipality lists the influx and high percentage of 'unemployed and unskilled indigent people' as its top strategic risk, adding significant pressure to service delivery (e.g., waste collection, education) and reducing tax revenue (TWK, 2015a, p. 35). The municipality aims to establish more permanent work through local economic development to reduce unemployment in winter and integrate people into the existing communities.

The people from other provinces who initially come for seasonal work on the fruit farms but then stay in the valley have mainly settled in the informal areas on the outskirts of the town, where public services, such as sanitation and solid waste management, are minimal or absent. There are strong contrasts between the different neighbourhoods of Grabouw in socio-economic characteristics, racial composition, levels of formalization and access to sanitation. The formal settlements are characterized by high living standards, while the informal settlements have low living standards, with substandard housing and poor access to municipal services. The situation has changed little since Peters and Wilson noted in 2007, 'There is a distortion in service provision [in Grabouw] based on race that reflects the apartheid history[…]. Apartheid planning, combined with the recent migration of people and, limited resources within the municipality, impacted on the extension of bulk infrastructure to previously un-serviced (and predominantly black) areas' (p. 290). The lower-income coloured and black households largely live in state-subsidized or informal houses. There are also big differences in the water supply and sanitation facilities: while some areas have private taps and toilets, others have communal taps and toilets, and in at least three neighbourhoods the facilities are so poor that 'people resort to using the surrounding bushy areas [for open defecation]' (p. 290). Residents in the low-income areas are pressuring the municipality for housing development and service delivery to redress the strong imbalance in living conditions. Grass-roots protests have emerged in the Elgin Valley, and across South Africa, as a key mechanism for this and are discussed below.

Housing construction and the formalization of informal settlements is one of the municipal responsibilities to redress the infrastructure inequalities created under apartheid, but housing development and the construction of basic water and sanitation infrastructure in Grabouw are not keeping up with its growth (TWK, 2015b, 2016; Western Cape Government, 2015). Newly constructed sewer networks have been connected to the existing municipal network, overburdening the main sewerage lines that convey wastewater to the Grabouw Wastewater Treatment Plant. According to municipal officials and the GWUA, the sewer system is frequently unable to convey all the wastewater, and as a result, considerable amounts of raw, untreated wastewater end up in the Swannies and Klipdrift Rivers (tributaries of the Palmiet River that flow through town). The Theewaterskloof Municipality indicates that it is trying to improve the downstream water quality, having upgraded the wastewater treatment plant, by being active in housing development, and by closely monitoring sewage spillages. The latter is also done by the GWUA to quickly identify any spills and lobby the municipality to fix blockages in sewer lines. The municipality explicitly points out the critical link between the farmers' labour strategies and the challenges related to service provision in Grabouw: the labour strategies of the commercial farmers endangers the municipality's financial viability and sustainability because the influx of seasonal workers, the low wages, and the seasonality of labour increases the burden on the provision of housing and basic services for people who are not able to pay for services (Baleta, 2015; TWK, 2012).

The violent mass demonstrations in the area are a sign of the tensions and dissatisfaction with the persistent inequalities in the Elgin Valley (Marais, 2014). Local political protests around service provision are not unique to the area and are a recurring phenomenon throughout the country. Attempts to analyze their occurrence in time and space on the basis of policy registration have limitations, as they lack specifications of the reasons for social mobilization or are defined by physical police intervention (Alexander, Runciman, & Maruping, 2016; Nyar & Wray, 2012). Protests are used to draw attention to the quality of service delivery and form an important element in grass-roots political strategies (Alexander, 2010; Booysen, 2007). Although detailed information on the frequency of local protests in (peri-)urban Grabouw is lacking, they have occurred at least annually in recent years. Marginalized groups of black and coloured residents organize themselves in grass-roots organizations such as the Grabouw Elgin Civic Association to demand their rights and challenge existing inequalities. Over the past few years, there have been various marches and protests in the Elgin Valley, which have blocked the national N2 highway and, on one occasion, caused the looting and burning down of the Grabouw traffic department office. An example of protests is the urban settlement of Siyanyanzela in Grabouw (Box 1). The main reasons for discontent and demonstrations are the shortcomings of service delivery, lack of housing, and minimum wages for farm workers. The protests aim to pressure the authorities to improve service provision (e.g., access to adequate sanitation infrastructure) and people's economic prospects. These emphasize very different understandings of what constitutes 'proper sanitation' and 'economic opportunities' than those emphasized by the commercial fruit farmers. Connected to the urban protests are conflicts within the agricultural sector about seasonality of work, the security of their contracts, housing and social exclusion (TWK, 2015c), resulting in wage strikes, labour conflicts and violence. Such farm workers' protests have also occurred at different places outside the Elgin Valley (Wilderman, 2015).

Low-income households thus have a very different understanding of the valley's hydrosocial territory and emphasize the structural inequalities in service delivery and economic opportunities that exist in the valley. Though they consider themselves dependent on the agricultural sector, they also recognize that it does not provide a stable income, proper housing or access to basic water and sanitation.

Box 1. Siyanyanzela urban settlement.

A settlement in Grabouw that has been of recurring interest in the media because of clashes between residents and law enforcement is Siyanyanzela ('forcefully' in Xhosa), which has grown to around 1000 informal homes. Residents have occupied vacant land here because the other low-income (informal) areas in Grabouw are overcrowded. They currently rely on makeshift water and electrical connections, there are no official collection services for household waste, and open defecation is common. Residents have repeatedly called on the municipality to provide running water, toilets and electricity, but the municipality condemns the 'illegal' land occupation and denies the residents access to basic services (Gontsana & Kelly, 2016; Kalipa, Mazantsi, & Lepule, 2016; Mnyakamak, 2017)

Urban water pollution threatens hydro-territorial order

The quality of water in the Palmiet River has considerably degraded over recent decades (River Health Programme, 2003, 2011). This section analyzes the nature and extent of water pollution in the Elgin Valley and shows that there are different perceptions of what is the core problem, which reflect differences in the understanding of how the valley's hydrosocial territory is and should be constituted. Low-income households emphasize the inadequate sanitation and waste management services in the areas where they live. Commercial fruit farmers foreground the potential threat of water pollution to the economic vitality of the valley.

As part of the research by Wessels (2016), the results of three monitoring programmes were combined to identify the sources of pollution, their severity and their fluctuations over time. The sampling points are both upstream and downstream of Grabouw and cover the tributaries flowing through the urban areas of Grabouw. The analysis shows a clear pollution hotspot of anthropogenic pollution in and around Grabouw. In the upper stretch of the Palmiet River, near the Eikenhof Dam, the water quality is generally good.[2] Moving downstream, it deteriorates, especially in summer, when much water is abstracted from the river for irrigation. *E. coli* levels in the mid-section of the Palmiet increase towards the end of summer (March and April). This indicates faecal contamination in the river, a health risk if transmitted to consumers of the irrigated produce (WHO, 2006). High concentrations of suspended solids and chemical oxygen demand in the Swannies and Klipdrift Rivers indicate domestic and industrial waste in the water. Further downstream, the water in the Kogelberg Dam is once again of good quality, as indicated by low levels of *E. coli* and chemical oxygen demand, illustrating the self-purification of the river and the dilution of the pollutants by inflows of relatively clean water from tributaries. The variation in water quality shows the risk of water pollution specifically for irrigated agriculture downstream of Grabouw during and towards the end of the water-scarce summer.

Based on the quantitative data as well as interviews with municipality officials and farmers and field visits with municipal workers, one of the primary sources of pollution is the discharge of untreated urban wastewater from low-income areas, such as Rooidakke and Pineview (Figure 2). The waste management problems in these areas are the result of personnel shortages, mixed waste streams, blurred district municipality functions and insufficient equipment (Delcarme, 2017). Other sources of pollution relate to the overloading of the sewer system and treatment plant, industrial effluent exceeding the discharge standards, direct industrial runoff into the environment, and wastewater discharge from workers' houses in farm compounds. The analysis shows that the wastewater treatment plant just downstream of Grabouw has been a source of pollution (Wessels, 2016), but the farmers, the GWUA and the Theewaterskloof Municipality indicate that the plant is functioning better since it was upgraded, although this cannot be verified by the water quality measurements, since there is limited post-upgrade data. Sewage overflows remain a common problem and are a result of solid waste causing blockages and overloading of the sewer system. As a result, raw wastewater sometimes ends up in the river and can account for a relatively high share of the total river discharge, especially during the drier summer months, when water is used for irrigation (Wessels, 2016).

The commercial farmers describe the water pollution issues in Grabouw as threatening their export markets. Access to clean water for irrigation is presented as a requirement for agricultural production and thus crucial for the economic vitality of the valley, on which many people depend. Residents in the low-income areas, however, see pollution from the urban areas as a result of the dire conditions in which they live and the limited economic opportunities in the valley. The labour strategies of the commercial farmers and structural imbalance in economic prospects are presented as main reasons for the water pollution challenges related to service provision in the low-income areas. The analysis of water quality (Wessels, 2016) finds that the direct application of river water to the crops would endanger access to export markets, particularly in the summer and in the area directly downstream of Grabouw. Elevated levels of pollution in the river, largely resulting from urban areas, thus threatens the dominant hydrosocial territory, structured around the interests of the commercial fruit farmers.

Water reforms to redress inequities and fruit farmers' defensive responses

Discussions and actions concerning water quality in the Elgin Valley do not result only from supposedly 'natural-technical' conditions but are embedded in political-institutional and social processes at various scales. In this section we analyze how farmers have dealt with the threat of losing control over water (quality) management in the context of post-apartheid water reforms that aimed to upset the status quo and redress long-standing racial, socio-economic and political inequalities. As we demonstrate below, fruit farmers actively operate to maintain control over the institutionally changing environment in a political-institutional context that is shifting as a result of different water reforms meant to decentralize and democratize decision making about water access and control. Besides political-institutional strategies to maintain and improve water quality in the river system, fruit farmers also take individual technical measures to protect the quality of their water sources and prevent losing access to export markets. Both are part of a strategy to protect their economic and political interests in the dominant hydro-territorial order.

The National Water Act of 1998 aims to ensure that water resources in South Africa are protected, used, developed, conserved, managed and controlled in a sustainable manner, and to explicitly redress apartheid-era water inequities, though the interpretation of what this means has evolved over time (Movik, 2014). The purpose of the act is to manage the nation's water resources, taking into account (among other things):

(a) meeting the basic human needs of present and future generations;
(b) promoting equitable access to water;
(c) redressing the results of past racial and gender discrimination; …
(g) protecting aquatic and associated ecosystems and their biological diversity;
(h) reducing and preventing pollution and degradation of water resources; …
(k) … and for this purpose, to establish suitable institutions and to ensure that they have appropriate community, racial and gender representation. (RSA, 1998, p. 18)

The National Water Act established CMAs as part of its strategy to retrench and contest existing inequalities through new democratizing institutional structures, yet two decades on structural inequalities still persist (Movik, Mehta, van Koppen, & Denby, 2016). CMAs were tasked with the development of catchment management strategies to protect, develop and control water resources. CMAs were on the one hand made independently responsible for securing the lawful use of water, but on the other remained under strict guidance of the Department of Water and Sanitation. Among other things, they are empowered to recover all reasonable costs from those responsible for illicit water pollution. For the Elgin Valley, the Breede-Overberg CMA became operational in 2007, and its management strategy was signed in effect in 2014. In May 2014, the Breede-Overberg CMA merged into BGCMA, as part of the process of reducing the number of CMAs from the original 19 to just 9. Due to the changes in its structure, the BGCMA lacked the capacity to actively engage with water issues in the Elgin Valley, resulting in a situation where, at the time of field work, no organization had the capacity to investigate and prosecute cases of unlawful water use (personal interview with a BGCMA official; Meissner, Funke, & Nortje, 2016).

Under the CMA, water user associations (WUAs) operate at a local level and are considered cooperatives of individual water users that undertake water-related activities for their mutual benefit. The existing irrigation organizations of white commercial farmers (irrigation boards) were seen as a starting point for the establishment of WUAs, while the legal guidelines on the transformation of irrigation boards into WUAs emphasized the need for the inclusion and representation of HDIs in these associations (DWA, 2000). Though these reforms aimed to also overturn power structures and give HDIs influence in water management decisions, in many instances powerful players used the WUAs to maintain the status quo. Studies from other sites in South Africa show that water struggles between commercial farmers and communities and decision making within WUAs tend to reflect existing power relations at a local level, since all the initiatives of new water users need to be discussed and agreed on in the WUAs, which are dominated by white commercial farmers and where emerging farmers from historically disadvantaged communities are still marginalized (Förster, Downsborough, & Chomba, 2017; Kemerink, Mendez, Ahlers, Wester, & Van der Zaag, 2013). The GWUA was based on the former Groenland Irrigation Board as part of the decentralization process instigated by the National Water Act. In practice, it functions as an advocacy group for commercial farmers' interests regarding water. The majority of the management committee of the GWUA consists of white commercial farmers, and interviews with farmers and GWUA representatives confirmed that the association is effective at representing their interests.

To secure access to good-quality water, commercial farmers employ both technical and institutional strategies, ranging from adjusted farming practices to collectively pressuring state authorities to act according to the economic interests of the valley. Although this is not their legal responsibility, the GWUA has appointed someone to actively monitor water quality in and around Grabouw. This person is a (part-time) commercial farmer, well known and well respected by the farmers, the municipality and the GWUA board. His activities are acknowledged by all to have had a significant impact on the quick resolution of sewage spills over the last years. By doing this, the GWUA monitors the sewage network in the urban areas of Grabouw and puts pressure on the municipality to fix sewage spills. The association is also planning to construct an artificial wetland in the

Palmiet River, downstream of the Klipdrift River, that will reduce contamination from the Swannies and Klipdrift Rivers, with the main objective of preventing pollution incidents affecting the irrigation practices of downstream farmers.

Because of the high level of dam development in the Elgin Valley, water can be stored and released in response to downstream demand. These dams are managed by the GWUA. In summer, water flows are often minimized to the legally required (environmental) flow. Thus, when sewage spills occur, the concentration of pollutants exceeds the local standards. When the river quality deteriorates because of a pollution event, it is possible to increase the discharge from the Eikenhof Dam and flush the river so that farmers pumping from the river are not too badly affected. These practices show that WUAs have allowed commercial farmers to maintain control over water flows and thereby at least partially mitigate water quality problems, without challenging the underlying structural issues and inequalities related to water access, control and decision making, or the existing labour regime. In other words, the technical-managerial measures of the GWUA reduce a complex situation that is co-constituted by social, natural and political processes to a mere biophysical problem that can be solved 'objectively'. The deeply political nature of the problem and the proposed solutions is concealed, leaving the hydro-territorial order intact and in line with commercial famers' interests. The democratization of water management, as intended by the water reforms, remains limited.

Besides trying to protect the quality of water sources through the WUA, farmers have several other strategies to secure their access to export markets. First, there are rumours that farmers have manipulated the water quality monitoring test results on which their accreditation and market access depend. In interviews it was suggested that farmers have been known to collect water from locations other than their own irrigation source or to order specific results from collaborating laboratories. The BGCMA reported that some commercial laboratories in the Breede-Gouritz area have lost their licences because of such incidents having been discovered. Nowadays, accreditation processes require external companies to do the sampling to minimize fraud. Second, pollution in the water source does not automatically end up in the apples and pears. Various risk-reduction layers exist that reduce the risk and minimize the adverse effects of using polluted water in crop production and its potential impact on health (and access to export markets). For example, in late summer farmers pump water from deep below the surface of one of the dams, as the water quality near the surface is often poor at that time of year. At the farm level, filtration systems are used in combination with localized irrigation systems (drip and micro-sprinklers). While these filtration systems were primarily installed to prevent clogging, they also help remove pathogens. On-farm water reservoirs, originally built for storage, also enhance pathogen removal and reduce the risk of water contamination. Lastly, the fruits that are harvested, packed and marketed go through chlorine drenches before leaving the packing facility, to eliminate any pathogens that may be on the surface of the fruits. Such measures effectively provide layers of risk reduction, and this study did not find any measurable risk to consumers (Wessels, 2016). In this way, the farmers' access to financial resources, technologies, institutions and water control allows them to maintain their economic-political position in the valley. Despite the political-institutional water reforms that aimed to upset the power balance, the way commercial fruit farmers in the Elgin Valley have dealt with these reforms is another example where in practice the reforms reasserted farmers' power. Through a combination of technological and institutional strategies, hydrosocial dynamics are maintained despite ongoing contestation.

Conclusions: overturning the hydro-territorial order?

In this article we have explored the encounter between divergent ideas and contestations around water and pollution flows in the Elgin Valley through the lens of hydrosocial territories. We have analyzed why and how relatively impoverished (coloured and black) urban dwellers of Grabouw continue to lack access to basic water and sanitation services, fair working conditions and economic opportunities alongside, and in contrast to, a vibrant commercial fruit sector dominated by white commercial farmers and large agricultural export companies.

Studying the case from a hydrosocial territories perspective puts the emphasis on processes of building, maintaining and protecting of territory, i.e., constructing convincing and dominant imaginary and socio-environmental materializations of spatially bounded heterogeneous water networks. The heterogeneity of these networks relates to the transformation of interlinked biophysical, technical, legal, institutional, cultural, discursive and symbolic arrangements in the aim of establishing and/or protecting order. Obscuring the political nature of these processes, among others, by rendering them technical, is part and parcel of successful territorialization.

Also, in the Elgin Valley the dominant hydrosocial territory is maintained by the entwinement of discursive, institutional and technical practices. The valley's welfare is *discursively* portrayed as dependent on fruit exports, to which water pollution poses a major threat. *Institutionally*, the WUA and the CMA, introduced as part of water reforms with the aim of advancing the power of HDIs, became dominated by commercial farmers and their interests. 'Inclusion-oriented' policies thus came to reinforce the dominant hydro-territorial order rather than opening it up (cf. Boelens et al., 2016). *Technically*, irrigation water quality is tackled mainly at the field level and by emergency responses to sewage spills, instigated by farmers' surveillance practices, bypassing the municipality. By presenting water quality as a 'practical challenge' farmers render technical a set of deeply political questions regarding skewed wealth distribution and development opportunities with a history of violent resource capture and political oppression along class and racial lines, which continue to be manifest in the living conditions of large groups of urban poor that contrast to the commercial interests of a relatively small group of high-tech commercial farmers. Moreover, certain elements of the associated political economy are backgrounded – notably problems related to living conditions for the seasonal workforce, who relocate to nearby informal settlements during the off-season. The lack of suitable infrastructure to serve those communities presents potential risks for public health and water quality, and associated threats to the economic activities in the valley. Although the precise source of pollution is not necessarily known, certain issues and possibilities are emphasized, and others minimized, reflecting the power dynamics of the valley.

In contrast to the dominant hydrosocial territoriality, grass-roots movements of urban low-income households emphasize the inequality in housing provision, municipal sanitation services and low-wage and seasonal farm jobs and stress the structural imbalance in economic opportunity and service delivery. Their protests remain largely reactive to immediate injustices, and they appear to be struggling to make inroads into the political process or to create counter-images of how the social-natural could be ordered (cf. Hommes et al., 2016). Their counter-narrative seemingly does not mobilize to advance an alternative territoriality that could claim and materialize institutional and symbolic spaces and that might politicize infrastructure investment priorities and recast what is of essential value for the valley. Though the dominant order is contested by different ideas of how the Elgin Valley and its water use practices are, and should be, constituted, these ideas have so far been incapable of upsetting the proverbial apple cart. The dominant understandings of water problems and solutions facilitate the stabilization of the political order and 'actively depoliticize … inequality, misrecognition, and political exclusion' (Boelens et al., 2016, p. 10).

Whereas most studies about rural–urban water struggles relate to water quantity issues with transfers from rural to urban spaces, this article presents a conflict over water quality in which not the urban but the farming community has strong control over water resources. Through our analysis of events, actions and representations of water quality issues, this case thus illustrates how urban and rural areas can be interlinked and be interdependent in very diverse ways. Passing through urban and rural areas in the Elgin Valley, water takes on (pollution) characteristics that link urban living conditions to commercial farming practices, spaces that are otherwise discursively and politically largely disconnected. Also, one and the same person may play different roles in the rural and the urban space. Many urban residents are dependent on income and employment opportunities from the agricultural sector, but at the same time they observe the farmers' labour strategies and the structural imbalance in economic prospects they face. When looking at cases of water pollution and the reuse of urban water in a rural setting, water flows and pollution interlink the rural and the urban in unexpected ways.

As shown in this article, a hydrosocial territories lens contributes to understanding the diversity and dynamics of actors' interests and how they transform interlinked socio-technical, political-institutional and discursive-symbolic arrangements to establish and maintain territorial order in relation to rural–urban water struggles, especially when these are shaped around water quality more than around quantity issues and when rural rather than urban actors are in a position of power.

Acknowledgments

We thank the International WaTERS Network for financially supporting this research. The guest editors of this special issue and two anonymous reviewers have contributed significantly to improving this article. We are grateful to Nick Parrott of TextualHealing.eu for his careful language editing of the text.

Notes

1. Black, coloured, Indian and white were the four official racial categories in South Africa during apartheid.
2. The quantitative analysis in Wessels (2016) uses a water quality classification (good, fair, poor) based on the National Water Act (DWA, 2013; RSA, 1998).

Disclosure statement

No potential conflict of interest was reported by the authors.

Funding

This work was supported by the International WaTERS research and training network SSHRC grant [890-2013-0020];

References

Alexander, P. (2010). Rebellion of the poor: South Africa's service delivery protests–A preliminary analysis. *Review of African Political Economy*, *37*(123), 25–40. doi:10.1080/03056241003637870

Alexander, P., Runciman, C., & Maruping, B. (2016). The use and abuse of police data in protest analysis South Africa's Incident Registration Information System (Iris). *South African Crime Quarterly*, *58*(1), 9–21. doi:10.17159/2413-3108/2016/v0n58a1513

Baleta, H. F. (2015). *The concept of shared risk in public and private sector water security: A case study of Grabouw and the Elgin Valley, Western Cape, South Africa*. Cape Town: University of Cape Town.

Bernstein, H. (2013). Commercial agriculture in South Africa since 1994: 'Natural, simply capitalism'. *Journal of Agrarian Change*, *13*(1), 23–46. doi:10.1111/joac.12011

BGCMA. (2017). *Catchment Management Strategy for the Breede-Gouritz Water Management Area*. Worcester: Breede-Gouritz Catchment Management Agency.

BOCMA. (2011). *Breede-Overberg Catchment Management Strategy*. Worcester: Breede-Overberg Catchment Management Agency.

Boelens, R., Hoogesteger, J., Swyngedouw, E., Vos, J., & Wester, P. (2016). Hydrosocial territories: A political ecology perspective. *Water International*, *41*(1), 1–14. doi:10.1080/02508060.2016.1134898

Booysen, S. (2007). With the ballot and the brick: The politics of attaining service delivery. *Progress in Development Studies*, *7*(1), 21–32. doi:10.1177/146499340600700103

Bosch, D. (2008). *A history of the palmiet river*. Grabouw: Groenland Water User Association.

Boudreaux, K. (2010). Land reform as social justice: The case of South Africa. *Economic Affairs*, *30*(1), 13–20. doi:10.1111/j.1468-0270.2009.01967.x

Delcarme, B. (2017). *Situational analysis of waste management practices of municipalities situated in the Breede-Gouritz Water Management Area*. Cape Town: Centre for Water and Sanitation Research.

Du Toit, A., Kruger, S., & Ponte, S. (2008). Deracializing exploitation? 'Black economic empowerment' in the South African wine industry. *Journal of Agrarian Change*, *8*(1), 6–32. doi:10.1111/j.1471-0366.2007.00161.x

DWA. (2000). *Guide on the transformation of irrigation boards and certain other boards into water user associations*. Pretoria: Department of Water Affairs.

DWA. (2013). *Revision of general authorisations in terms of section 39 of the national water act, 1998* ((Act No. 36 of 1998), No. 665. Government Gazette). Pretoria: Department of Water Affairs

DWAF. (2004). *Breede water management area - internal strategic perspective*. Pretoria: Department of Water Affairs and Forestry.

Ewert, J., & Du Toit, A. (2005). A deepening divide in the countryside: Restructuring and rural livelihoods in the South African wine industry. *Journal of Southern African Studies*, *31*(2), 315–332. doi:10.1080/03057070500109524

Ferguson, J. (1994). The anti-politics machine. *The Ecologist*, *24*(5).

Förster, J. J., Downsborough, L., & Chomba, M. J. (2017). When policy hits practice: Structure, agency, and power in South African water governance. *Society & Natural Resources, 30*(4), 521–536. doi:10.1080/08941920.2016.1268658

GlobalGAP. (2016). *Quality Management System Checklist*. Cologne: GlobalGAP.

Gontsana, M., & Kelly, S. (2016). *Grabouw: "We are sitting here, lost"*. Retrieved from http://www.groundup.org.za/article/grabouw-we-are-sitting-here-lost/

GWUA. (2014). *Constitution of the Groenland Water Users Association*. Grabouw: Groenland Water User Association.

Hall, R. (2004). A political economy of land reform in South Africa. *Review of African Political Economy, 31*(100), 213–227. doi:10.1080/0305624042000262257

Haysom, G. G. D. (2007). *Project proposal for transforming Grabouw, Western Cape, into a sustainable community*. Stellenbosch: Stellenbosch University.

Hommes, L., Boelens, R., Harris, L., & Veldwisch, G. J. (2019). Rural-urban water struggles: Urbanizing hydrosocial territories and evolving connections, discourses and identities. *Water International, 44*(2), 81–94. doi:10.1080/02508060.2019.1583311

Hommes, L., Boelens, R., & Maat, H. (2016). Contested hydrosocial territories and disputed water governance: Struggles and competing claims over the Ilisu Dam development in south-eastern Turkey. *Geoforum, 71*, 9–20. doi:10.1016/j.geoforum.2016.02.015

Hoogesteger, J., Boelens, R., & Baud, M. (2016). Territorial pluralism: Water users' multi-scalar struggles against state ordering in Ecuador's highlands. *Water International, 41*(1), 91–106. doi:10.1080/02508060.2016.1130910

Hortgro. (2016). *Key deciduous fruit statistics 2016*. Paarl: Hortgro.

Kalipa, S., Mazantsi, S., & Lepule, T. (2016). N2 closed due to Grabouw protest mayhem. *Cape Argus, Cape Times and Daily Voice*. Retrieved from http://www.iol.co.za/news/crime-courts/n2-closed-due-to-grabouw-protest-mayhem—pics-2020223

Kemerink, J., Mendez, L., Ahlers, R., Wester, P., & Van der Zaag, P. (2013). The question of inclusion and representation in rural South Africa: Challenging the concept of water user associations as a vehicle for transformation. *Water Policy, 15*(2), 243–257. doi:10.2166/wp.2012.127

Lahiff, E., & Cousins, B. (2005). Smallholder agriculture and land reform in South Africa. *IDS Bulletin, 36*(2), 127–131.

Leibbrandt, M., & Levinsohn, J. (2011). *Fifteen years on: Household incomes in South Africa*. Cambridge: National Bureau of Economic Research.

Li, T. M. (2007). *The will to improve: Governmentality, development, and the practice of politics*. Durham, NC: Duke University Press.

Linton, J., & Budds, J. (2014). The hydrosocial cycle: Defining and mobilizing a relational-dialectical approach to water. *Geoforum, 57*, 170–180. doi:10.1016/j.geoforum.2013.10.008

Mabaya, E., Tihanyi, K., & Karaan, M. (2011). *Case studies of emerging farmers and agribusinesses in South Africa*. Stellenbosch: African Sun Media.

Marais, P. (2014). *What lies behind Grabouw riots?* Retrieved from https://www.iol.co.za/news/what-lies-behind-grabouw-riots-1753176

Meissner, R., Funke, N., & Nortje, K. (2016). The politics of establishing catchment management agencies in South Africa: The case of the breede-overberg catchment management agency. *Ecology and Society, 21*(3). doi:10.5751/ES-08417-210326

Mnyakamak, M. (2017, June 8). *Land occupiers in Grabouw demand services*. Retrieved from http://www.groundup.org.za/article/land-occupiers-grabouw-demand-services/

Movik, S. (2014). A fair share? Perceptions of justice in South Africa's water allocation reform policy. *Geoforum, 54*, 187–195. doi:10.1016/j.geoforum.2013.03.003

Movik, S., Mehta, L., van Koppen, B., & Denby, K. (2016). Emergence, interpretations and translations of IWRM in South Africa. *Water Alternatives, 9*(3), 456–472.

News24. (2014). *E coli in Grabouw water*. Retrieved from http://www.news24.com/Green/News/E-coli-in-Grabouw-water-20140606

Nyar, A., & Wray, C. (2012). Understanding protest action: Some data collection challenges for South Africa. *Transformation: Critical Perspectives on Southern Africa, 80*(1), 22–43. doi:10.1353/trn.2012.0050

Peters, K., & Oldfield, S. (2005). *The paradox of 'free basic water'and cost recovery in Grabouw: Increasing household debt and municipal financial loss. Urban Forum, 16*(4), 313–335. doi:10.1007/s12132-005-0009-9

Peters, K., & Wilson, Z. (2007). Chapter 12: Grabouw Case Study. In *Second order water scarcity in Southern Africa*. London: Department for International Development.

River Health Programme. (2003). *State-of-rivers report: Diep, Hout Bay, Lourens and Palmiet river systems*. Pretoria: Department of Water Affairs and Forestry.

River Health Programme. (2011). *Rivers of the breede water management area*. Pretoria: Department of Water Affairs and Forestry.

RSA. (1998). *National water act (Act No. 36 of 1998)*. Cape Town: Republic of South Africa.

StatsSA. (2011). *Grabouw Census 2011*. Pretoria: Statistics South Africa.

Swyngedouw, E. (1999). Modernity and hybridity: Nature, regeneracionismo, and the production of the Spanish waterscape, 1890–1930. *Annals of the Association of American Geographers, 89*(3), 443–465. doi:10.1111/0004-5608.00157

Swyngedouw, E., & Boelens, R. (2018). "… And not a single injustice remains": Hydro-territorial colonization and techno-political transformations in Spain. In R. Boelens (Ed.), *Water justice* (pp. 115–133). Cambridge: Cambridge University Press.

Swyngedouw, E., & Williams, J. (2016). From Spain's hydro-deadlock to the desalination fix. *Water International, 41*(1), 54–73. doi:10.1080/02508060.2016.1107705

Thompson, L. M. (1990). *A history of South Africa*. New Haven, CT: Yale University Press.

TWK. (2012). *3rd Generation IDP 2012–2017: First revision 2013/2014*. Caledon: Theewaterskloof Municipality.

TWK. (2015a). *Annual report 2013/2014*. Caledon: Theewaterskloof Municipality.

TWK. (2015b). *Integrated development plan 2012–2017: 3rd annual revision 2015/2016*. Caledon: Theewaterskloof Municipality.

TWK. (2015c). *Theewaterskloof's economy - an overview of the theewaterskloof municipal economy and economic development strategy 2015*. Caledon: Theewaterskloof Municipality.

TWK. (2016). *Integrated development plan 2012–2017: 4th annual review 2016/2017*. Caledon: Theewaterskloof Municipality.

Vink, N., & Kirsten, J. (2003). Agriculture in the national economy. In *The challenge of change: Agriculture, land and the South African economy* (pp. 3–20). Pietermaritzburg: University of Natal Press.

Wessels, M. T. (2016). *Understanding urban-rural linkages in water reuse for irrigated agriculture in the Elgin Valley, South Africa*. Wageningen: Wageningen University and Research.

Western Cape Government. (2015). *Socio-economic profile theewaterskloof municipality*. Cape Town: Western Cape Government.

WHO. (2006). *Guidelines for the safe use of wastewater, excreta and greywater: Wastewater use in agriculture* (Vol. 2). Geneva: World Health Organization.

Wilderman, J. (2015). *Farm worker uprising in the Western Cape: A case study of protest, organising, and collective action*. Johannesburg: University of Witwatersrand.

Hydrosocial territories, agro-export and water scarcity: capitalist territorial transformations and water governance in Peru's coastal valleys

Gerardo Damonte and Rutgerd Boelens

ABSTRACT

In recent decades, an agro-export boom has deeply transformed Peru's coastal valleys, resulting in dramatic territorial changes and social inequality in the Ica Valley. This article explains how politico-economic and socio-institutional forces have triggered the emergence of a new 'hydrosocial territory', transforming the Ica Valley into a virtual-water extraction zone that produces luxury export crops for the North and China. In addition, it shows how these territorial reconfigurations have led to ecological damage, water scarcity and increasing rural–urban inequality sustained by a hegemonic development discourse that supports agribusiness-elite territorial dominance and discourages social unrest.

Introduction

With the boom of agro-export business on the Peruvian coast, the agricultural frontier has rapidly expanded over the last decades. Due to large water infrastructure projects and intensive groundwater use, farming is no longer restricted to the valleys but overruns the arid zones in their surroundings. These changes have happened in the context of new power relationships among agricultural producers living in coastal valleys and, in many cases, have reinforced the groundwater scarcity that affects both rural and urban areas. The agricultural expansion generates employment for poor labourers (albeit with low wages and long, harsh working days), but also increases social inequality and depletion and contamination of surface and groundwater resources (Damonte, Pacheco, & Grados, 2014; Oré & Damonte, 2014).

The Ica Valley, in the lower Ica River basin, 300 km south of Lima, is a paradigmatic illustration. The river's headwaters are in the upper basin, on the western slopes of the Andes mountain range. In the upper river valley, the basin has been expanded by including lagoons from the neighbouring region of Huancavelica through the construction of inter-basin water transfer infrastructure: the Choclococha I Integrated System,

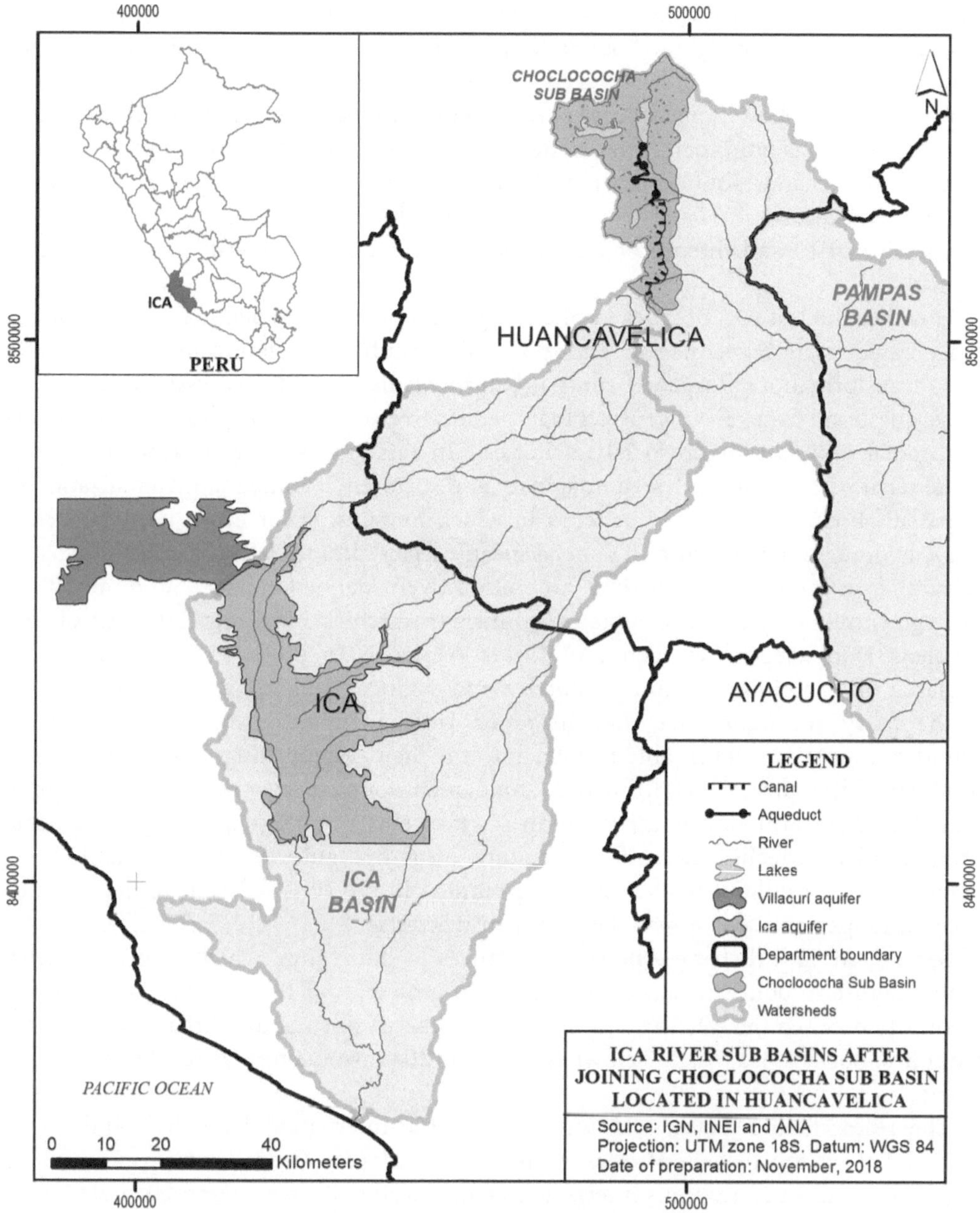

Figure 1. Ica River sub-basins after joining the Choclococha sub-basin in Huancavelica.
Source: IGN, INEI and ANA.

constructed in the mid-twentieth century. The lower valley has two aquifers: the Ica aquifer and the Villacurí and Lanchas aquifer (Figure 1).

In the 1990s, the government's policy was to appeal to private investments for agro-export expansion. This policy attracted domestic and foreign capital, introducing new export crops. With increased demand for asparagus on the international market, by the end of the 1990s the Ica Valley was one of the country's main agro-export zones. The agro-export boom brought, along with population growth, increasing demand for groundwater.

As this article shows, this process affects urban water supply in the city of Ica, which especially depends on underground water resources and infrastructure since, due to the heightened demand, aquifers have been over-pumped, triggering scenarios of water scarcity (Cárdenas, 2012; Marshall, 2014; Oré, Bayer, Chiong, & Rendón, 2012; Progressio, 2010) and intensifying dynamics of competition for water access, use and control.

How has the last two decades' capitalist agro-export boom redefined the Ica Valley's water-based material-economic and political-institutional relationships? How have these transformations deepened rural–urban inequality in water access? We approach these questions from a political ecology perspective, in particular elaborating on the conceptual notion of hydrosocial territories.[1] In this analysis, we understand hydrosocial territories as 'the contested imaginary and socio-environmental materialization of a spatially bound multi-scalar network in which humans, water flows, ecological relations, hydraulic infrastructure, financial means, legal-administrative arrangements and cultural institutions and practices are interactively defined, aligned and mobilized through epistemological belief systems, political hierarchies and naturalizing discourses' (Boelens, Hoogesteger, Swyngedouw, Vos, & Wester, 2016, p. 2).

Based on examining techno-productive and political-geographical reconfigurations in the valley, we argue, first, that the power-laden imaginaries and actions of agro-export networks have been able to filter into the State, establishing a dominant water-production discourse, which formed the fundament for shaping a new hydro-territorial grid. This transformation is expressed in physical terms by the expanding agricultural frontier based on techno-scientific agriculture, mega-hydraulic development and water-grabbing, and is politically and socially grounded in the consolidation of new institutional arrangements and a new development discourse.

Second, we find that these power imbalances in the region, which favour capitalist mono-culture, produce unsustainable, unequal growth, and thereby deepen territorial inequality between the wealthy and the poor and between rural and urban dwellers (see Hommes, Boelens, Harris, & Veldwisch, 2019, in the introductory article to this special issue).

The research took place, and continues, in the international research and action carried out by the Justicia Hídrica/Water Justice alliance (http://www.justiciahidrica.org). The project's continuous interaction among academic and vernacular/grass-roots-knowledge actors has enabled cross-breeding of the research findings. Methods were based on sequential use of qualitative research methods. The first stage was a systematic literature review, covering the theoretical and historical literature and case studies. This was complemented by compiling economic and geographical data. The second stage was ethnographic data collection in 2013, with follow-up in 2014–2017. Various tools were applied: interviews, participant observation and participatory mapping. Specifically, we conducted debate and observation sessions, formal and informal meetings of users, and in-depth semi-structured interviews with key regional stakeholders, such as agro-exporters, medium-scale producers and government officials. The third stage systematically organized the information in reports and maps that were presented and discussed by the project's interdisciplinary team during regional workshops. Finally, information was validated with local and regional stakeholders in Ica through participatory workshops, and compared and triangulated with other case study findings in national and international seminars.

This article is structured as follows. In the next section we present the conceptual notions. Thereafter we describe the agro-export boom in Ica and present details about the Ica Valley. Then we analyze the production of water inequality between, on the one hand, agro-export rural zones and, on the other, smallholder rural areas and poor neighbourhoods in Ica City. We end with a scrutiny of the dominant discourses and the conclusions: Peru's agribusiness elites instrumentalize water governance to suit their interests and accordingly shape Ica's hydrosocial territory, so that water transfers make the water flow in the direction of power.

Conceptual notions

Since the classic book by Wittfogel (1957), which argues that societies depending on large-scale irrigation infrastructure require centralized water control, which leads to governmental despotism, an abundant literature has been generated regarding the multiple forms interrelating water control and societal power. This shows that water-based forms of power and governmentality endeavours (Foucault, 1991 [1978]) can be located not only in the State and its administrative apparatus but also in societal stakeholders who, by various means, get control over water, its development, or its service provision (Bakker, 2010; Boelens, 2015; Harris & Alatout, 2010; Menga & Swyngedouw, 2018; Swyngedouw, 2009). This power can come from controlling the water resources, such as rivers and springs, and hydraulic infrastructure, such as major dams or canals (Duarte-Abadía, Boelens, & Roa-Avendaño, 2015; Hommes, Boelens, & Maat, 2016), or by deploying institutional, symbolic and discursive power forms, embedded in and sustaining dominant knowledge or policies that seek to subordinate local water knowledge and management (Boelens & Vos, 2012; Orlove & Caton, 2010). Water power can also be generated by changing property regimes or provision services, such as by nationalizing or privatizing water resources, water rights, and/or water services. For instance, when water is assigned a monetary value and water rights are made transferable, water becomes a commodity that may be exclusively accumulated by stakeholders with large economic power (Bakker, 2010; Duarte-Abadía & Boelens, 2016; Ioris, 2016). Power-through-water is often generated through a combination of mechanisms. For example, developing infrastructure and discourses on water efficiency can facilitate water-grabbing by national or global elites (Swyngedouw, 2009). In the case of groundwater, such processes have involved pumping out aquifers for large-scale agriculture, particularly in arid or semi-arid zones (Oré & Damonte, 2014; Wester & Hoogesteger, 2011).

This dialectical relationship between water and society has been approached through, for instance, the concept of the hydrosocial cycle. It refers to the ongoing interaction between the hydrological cycle and social developments: the complex, diachronic manner in which human action (with its inherent forms of culture and power) implicates and reconfigures water's natural cycle, and vice versa (Boelens, 2014; Budds, 2008; Linton & Budds, 2014). Closely related concepts revolve around the notion of waterscape, which reflects the social-natural networked connections made by hydrosocial cycle transformations in particular techno-political contexts (Baviskar, 2007; Budds & Hinojosa, 2012; Swyngedouw, 2009).

These notions manifest how territories are given shape as historical socio-physical constructs. Societal stakeholders construct, on the basis of their territorial visions, interests and narratives, the boundaries, contents, meanings and relationships of a given territory. Therefore, a territory's physical conformation is mediated by the relations of power and conflict among different territorial visions (Damonte, 2011; Elden, 2010; Harris & Alatout, 2010; Lefebvre, 1991).

The notion of hydrosocial territories integrates technical-physical, social and natural space and relationships, building on how people – in divergent ways – conceptualize and materialize water uses and plan to be, or are, connected by water flows and hydraulics. Hydrosocial territories imply imagined and actual boundaries: physical and social limits that are imposed and redefine the contexts of struggles for water power. It follows that hydrosocial territories (imagined, planned or materialized) have disputed functions, values and meanings (Hoogesteger, Boelens, & Baud, 2016; Rodriguez-de-Francisco & Boelens, 2016; Seemann, 2016). Their imagination, design and efforts towards implementation define processes of inclusion and exclusion, development and marginalization, and the distribution of benefits and burdens that affect different groups of people in distinct ways. 'Therefore, the question of how, by which actors, through which strategies and with what interests and consequences the "natural" and the "social" boundaries of hydrosocial territories are conceptualized and materialized through interlinked natural, social and technological elements, is fundamental' (Boelens et al., 2016, p. 3).

The concept of hydrosocial territories is especially suitable for a multidimensional analysis of the interplay between territorial and social transformations. As Swyngedouw and Boelens (2018) and Hommes and Boelens (2017) state, 'hydrosocial territory' presents a further elaboration compared to the above-mentioned concepts (hydrosocial networks, waterscape and hydrosocial cycle), on the one hand, and literature on territories, on the other. 'Most territory literature does not fundamentally theorize the particularities and fluid properties of technology (materialized and embedded norms) in shaping geographies (e.g., the (de)politicization and "moralization" of technology). Most waterscape literature focuses particularly on the hegemonic structures and discourses that drive (and follow from) socio-natural reconfiguration' (Swyngedouw & Boelens, 2018, p. 129). They continue explaining that (while maintaining the central notion that water flows, technologies, institutions and power structures are arts of statecraft and shape multi-scalar political geography) 'hydrosocial territory conceptual innovation stresses the resulting diversity in terms of overlapping, simultaneously existing hydroterritorial regimes and imaginaries (in one and the same geo-political location, and with unequal power)' (p. 130). Furthermore, the hydrosocial territories notion gives explicit attention to actor-oriented approaches and interface concepts (see e.g. Long, 2004). 'The concept also explicitly deals with modernist territorial recognition politics: the tactics of "recognizing" and "incorporating" local territorialities (integrating local norms, practices, and discourses into mainstream government rationality and its spatial/political organization) make state/formal and local/vernacular modes of territorial ordering depend on each other in complicated (and often confrontational) ways' (Swyngedouw & Boelens, 2018, p. 130).

The relationships between State, society and market forces co-define and mediate these socio-economic and techno-political structures, for instance through institutionalized administrative and technological practices to use and access water and territory (e.g., Harris & Alatout, 2010; Hommes et al., 2016). Here, the State can be portrayed as a largely internalized meta-organization designed and conceived to regulate, monitor, contain and normalize the strategies and interests of the organizations and interest groups comprising segments of society (Chatterjee, 2007; Corbridge, Wiliams, Srivastava, & Véron, 2005; Migdal, 1988).

In this sense, the State is not separate from society but a contingent arena where society's different organizations and actors – private and collective, local and transnational – vie over their interests, commonly under highly unequal market and power relationships (Joseph & Nugent, 1994; Li, 2005). As Bourdieu (1998) observed, the State exists in objective reality, as a set of institutions, but also in the way people perceive the State, 'in the minds of the people'. As a result, the State is an ambiguous reality, dependent on dominant forces in society and at the same time a battleground. Therefore, the State's capacity to exercise power and design, plan, implement or transform hydrosocial territories is contingent upon the balance of localized forces in and outside the State. This balance is also conditioned upon the government's institutional structure and procedures, and its relationship with local and transnational normative structures that impinge upon water control.

In imagining, building and contesting hydrosocial territories, the institutions (defined as rules or conventions grouping and coordinating legitimized social practices – Douglas, 1986) are key elements of the grid and the process and may be formal or informal. Here, as follows from the above, the lines dividing formal (official) from informal (unofficial) are not sharp; formal/informal is an institutional continuum, where stakeholders choose a course of action combining normative and customary, moral or traditional resources.

This article aims to contribute to the emergent literature on hydrosocial territories by linking multidimensional transformations with increasing rural–urban inequality. In particular, it highlights how development discourses are used to legitimize both territorial transformations and increasing inequality. The article shows that powerful development discourses can lay the foundations for hegemonic territorial dominance and sheer 'acceptance' of inequality. More specifically, it focuses on analyzing how the agro-export political project has managed to redefine hydrosocial territories in the coastal region of Ica, Peru. It thereby contests and subordinates diverse historically rooted peasant and indigenous territorialities, and also actively impedes imaginaries and materializations of hydrosocial territory that would be based on inclusive food security and sovereignty, extending even beyond Ica's city and valley to include the wider surroundings, such as those of the crucially important Lima region.

The emergence of a hydrosocial territory: boom of the agro-export economic model and political project

Since 1990s, the Peruvian neoliberal development project has privileged agro-export production based on water extraction as a way to boost economic growth. Under this political project, the state and agribusiness elites have consolidated a new agro-export economic model in the region, which in turn has produced a new hydrosocial territory.

The agro-export economic model as a political project

In Ica, a new hydrosocial territory has been gradually constituted based on an economic model of agro-exports. It inter-related three phenomena: imposition of a neoliberal project favouring large landholding; production of new high-yielding (and water-intensive) crops responding to worldwide demand; and concentration of water access rights in large production units producing for the global market. In this political-economic framework, valleys on the coast offer opportunities for large-scale development of crops enabled through mega-hydraulic projects.

The Ica Valley has become a particularly popular setting to grow export crops for several reasons. Ica is near the country's main highway network, and export harbours. The valley's temperate climate has no extreme temperatures, and there is enough sunlight for a variety of crops. However, the Fujimori government identified constraints on agro-export development. Small family farms were regarded as inefficient in producing for a global market requiring high quality, massive volume and constant production. Therefore, the government concentrated on appealing to and fostering the formation of export companies with access to large land areas. Mass-production of crops mainly for export has arisen next to traditional cotton. The model's economic success is intimately linked to the increased worldwide demand for fresh crop luxury foods, which goes along with steeply increased demand for resources (water and land) to grow these products. The main crops produced by the agribusiness are asparagus (whose harvest area has increased from 411 ha in 1991 to 10,400 ha in 2011) and table grapes (whose harvest area increases from 3000 ha in 1994 to 5082 ha in 2011) (Autoridad Nacional del Agua [ANA], 2018). The export destination for the table grapes has changed from Europe and the US to China because of less stringent agrochemical residue norms.

The agro-export boom in the Ica Valley has had profound consequences in water scarcity. The new export crops demand huge amounts of water, which becomes embedded in the exported produce as so-called 'virtual water' (Allan, 2003; Vos & Hinojosa, 2016). In an arid climate such as Ica's, expanding the agricultural frontier requires access to new water sources. In this scenario, agro-exporters are the ones with the necessary capital to access groundwater, which is the only additional high-quality water source available at this time. The National Water Authority (ANA) reports that in 2009 the exploitation of the aquifers surpassed the reserve volume by 125% and in 2013–2014 by 171% (Muñoz & Zuñiga, 2018, p. 114).

The agro-export model has concentrated resources through competition for water (Damonte et al., 2014; Muñoz, Navas, & Milla, 2014). Agro-exporters have accessed land in the valley by purchasing small farms and by State award of government-owned land considered unproductive and without official claims to property. The great majority of agro-exporters have accessed groundwater, multiplying the number of wells over the last few years. Their greater capacity to control irrigation with good-quality groundwater has been fundamental to agro-export success while depleting the aquifer, as shown by the ANA management plan.[2] In this context, State and local elites pushed landscape, institutional and political transformations that would 'solve water scarcity' and in turn defined the construction of a new hydro-territory.

Increasing water demands and landscape transformations in the Ica Valley

In the Ica region there are two mayor agricultural zones: the Ica Valley, fed by the Ica River, the Ica aquifer and water transferred from the Huancavelica highlands; and the Villacurí Pampa, a desert zone fed by the Villacuri aquifer.

The Ica River and aquifer feed some 26,000 ha of irrigated land in the Ica Valley, while also providing water to the city of Ica and surrounding villages. In total some 860 groundwater wells are installed, of which only 249 have a concession (ANA, 2018). In overall terms, three distinct types of users compete for the water resources: the smallholders at the bottom of the valley, the urban users, and the agro-export companies in the lower part of the valley.

The about 8000 smallholders have around 16,000 ha of irrigated land (62% of the total irrigated land). Two main irrigation systems provide surface water during the irrigation season: the La Achirana system (on the left bank of the Ica River) and the right bank irrigation canals. The smallholders are organized in two water users' associations: Junta de Usuarios la Achirana y Santiago de Chocorvos (La Achirana) and Junta de Usuarios del Rio Ica (JUDRI) (Ore, 2006).

Urban population has increased significantly in the last decades mainly due to immigration. The main provider of drinking water is the Municipal Drinking Water Company of Ica (EMAPICA); there are also several community-based providers of drinking water to urban families not connected to the EMAPICA network.

The third type of users are the agro-export companies, which have steadily increased their land and water access. Some 10 major agro-export companies have in total about 10,000 ha (38%); Beta with 2146 ha and Agrokasa with 1270 ha being the two largest. These agribusiness companies use groundwater for irrigation. However, some of the wells are shallow wells located very close to the river, thus actually pumping river water. In addition, several companies got access to smallholders' wells. They invaded the land of the smallholders by leasing or buying their land and also took possession of the surrounding deserts, pumping water from newly drilled wells (Cárdenas, 2012; Hendriks & Boelens, 2016).

The agro-export expansion has resulted in profound ecological and landscape transformations. The aquifer level has descended alarmingly, on average some 0.8 metres per year, drying up the shallowest wells, generally belonging to small farmers in other parts of the watershed, and ruining groundwater quality through salinization (Cárdenas, 2012; Oré et al., 2012). These changes prompted 'water bans' by the government in 2012 (ANA, 2018), which however have not prevented agro-export companies from continuing to drill new wells.

Intensifying agricultural production has also caused changes in land use, with large areas of monocrops (Marshall, 2014), transforming Ica's landscape features and territorial boundaries. The agricultural frontier has extended along the coastal desert, over-exploiting groundwater resources even more. And the new infrastructure to access underground water has changed the landscape beyond the agro-export companies' land.

The agro-export political project and state institutional transformations

The agro-export boom has also importantly changed the institutional grid of the Ica hydrosocial territory. Up to the turn of the century, State water management regulations did not cover the use of groundwater in Ica. Farmers dug wells mostly to complement their access to the volume of surface water. Land tenure above the aquifer automatically granted farmers rights to access groundwater if there was a well in use or when they had the financial capacity to make one. It was an informal institution. With the agro-export boom, the State started developing a legal framework to formalize the use of groundwater. A legislative decree granted protection for those users who invested in obtaining groundwater at their own cost and risk. This aimed to formalize the informal use of groundwater by companies. In 2009, the Water Resource Law, creating the National Water Authority (ANA), inaugurated a new institutional framework based on the paradigm of Integrated Water Resource Management (IWRM). This change, among other things, attempted to generate participatory mechanisms coordinated by the government, as well as de-concentrating public water management by forming decentralized provincial and local ANA offices. This new regulatory framework for groundwater, to be implemented by the staff of local water authorities (*autoridades locales del agua*), had little or no effect on slowing down capitalist water extraction practices.

Along with changes in the State institutional framework, water user organizations in Ica also changed. The most significant transformation was that the agro-export companies organized themselves into groundwater users' associations: JUASVI (Junta de Usuarios de Agua Subterránea del Valle de Ica) in 2009, which in 2014 had 59 members, and JURS (Junta de Usuarios de Agua Subterránea del Rio Seco). Through these organizations agro-exports could self-regulate extraction from the aquifer. This is especially important because it groups the region's new elites who are pumping this water. Interestingly, the existing surface water associations JUDRI and Junta de Usuarios de la Achirana did not oppose but even engaged with the powerful JUASVI.

These institutional transformations have consolidated the agribusiness elite's territorial dominance. The new agribusiness elite's groundwater users' associations provided institutional representatives to state agencies and other water associations. This way, these organizations have been able to avoid state regulation of the wells on agro-export companies' property and established political relations with other water associations to consolidate their privileged access to water resources.

The agro-export political project and discursive transformations

The agro-exporters are a heterogeneous group, united by shared interests in maintaining their market linkages and resource access – particularly access to groundwater. This group has also become the dominant actor in both economic and political terms in the Ica Valley. Their political power has been boosted by their capacity to establish the dominant development discourse and permeate State institutions (Damonte & Gonzales, 2018; Damonte, Gonzales, & Lahud, 2016). In their discourse, agro-exporters portray themselves as the drivers of development in Ica, refloating the local economy with their greater technical knowledge and entrepreneurial capacity. Agro-exporters' representatives all point out that their companies create jobs, pay taxes, help improve transport and irrigation, use high-tech irrigation, and monitor efficient use of water resources.[3]

Here, the idea of efficiency in water management is central to agro-export discourse (Boelens & Vos, 2012): water use is more efficient when less water is needed to produce a high-quality crop. This production efficiency can result in economic efficiency if the crop is marketed for a significant profit. Thus, the discourse goes, agro-exporters, who use efficient irrigation technology, must have privileged access to water (Oré & Damonte, 2014; Oré & Muñoz, 2018; Urteaga, 2010). However, policies pursuing water efficiency ignore the social impacts of granting privileged water access to a single production sector. In Ica, these water policies and market forces are generating increasing inequality in water access between users considered efficient or inefficient, such as small farmers who cannot afford to implement new, high-tech irrigation infrastructure (Oré & Damonte, 2014; Oré & Muñoz, 2018).

Despite its socio-ecological impacts, the agro-export discourse has few critics in the Ica Valley. Government officials feel that agro-exporters have greater financial and technical capacity than other stakeholders, so they accept agro-exporters' authority to manage water. Even small and medium farmers, who rent their land or sell their water to agro-exporters, often feel that the agro-export authority is legitimate. As an old smallholder, who has rented his land to a company, asserted: "They [agro-exporters] know how to manage water, they have technology.… It is like before the land reform, the *hacendados* [landlords] had technology and money to get good production.… We are poor, we do not have the means to crop high-quality grapes for export."

This can be explained by three reasons. First, several societal stakeholders in Ica have positive (often romanticized) memories of the hacienda period prior to the 1969 agrarian reform. They view some agro-exporters as 'good bosses', which shows that old power discourses have survived, incrusted in local residents' social life (Damonte et al., 2014).

Further, most local people have not (yet) realized the damage that depleting their aquifer can cause, not only for users of groundwater. The connection with surface water flows is 'invisible'. It is not evident for irrigators using surface water that export companies are competing with them for their water (cf. De Bont, Veldwisch, Komakech, & Vos, 2016). As a member of JURLAH declared:

> They [companies] use only groundwater, we use river water, from La Achirana channel, we do not compete with each other.… They know how to manage groundwater, they have well-educated engineers, we know them, sometimes they come and help us with water technical problems.… They have brought development to Ica.

Next, those people who do feel the consequences of governmental water-extractive policies and agro-export companies' water-predatory practices are the ones who lack voice or remain far away: for instance, local smallholders whose shallow wells (and corresponding livelihoods) have dried out, and Andean communities that suffer from Peru's aggressive mega-hydraulic projects diverting water from the (Amazon-)Andean watershed to the coast (Hoogesteger & Verzijl, 2015; Urteaga, 2014; Verzijl & Guerrero-Quispe, 2013).

The discursive practices are very important to the emergence of a new hydrosocial territory. The state and agribusiness elites have been able to legitimize agro-export's control over water resources by presenting the agro-export model as the only development model suitable to the region. This legitimacy has cemented the agro-export elite's capacity to push forward territorial transformations to facilitate its access to water without strong opposition.

Urban–rural dynamics and inequality in the realm of agro-export territorial transformation

The rapid growth of the city of Ica and the expansion of agro-economic activities in the region of Ica has put great pressure on water resources: the main water demand comes from agriculture and urban water consumption. The region's population grew from 447,000 to 764,000 between 1981 and 2012, and urban dwellers account for almost 80%. Most new residents were immigrants from the neighbouring Andean regions of Huancavelica and Ayacucho, and were displaced from their homes by poverty and violence (Oré, 2006). They settled in the poor urban and peri-urban areas.

Natural conditions are not favourable to fulfil the increasing water demand in the Ica River basin. The seasonal Ica River's flow depends on rainfall in the upper watershed: heavy rains produce water overflows, but the river dries when the rainy season ends, producing water shortage. Therefore, water provision depends on how water is managed. During the dry season, agriculture and the city rely on regulated water flows from reservoirs in the highlands and, especially, from underground water that comes from the Ica and Villacuri aquifers. Thus, Ica requires a public policy to ensure water access during the dry season and prevent floods.

Since the 1940s, EMAPICA has developed a water supply network that interconnects most urban sectors in the city of Ica and some in the neighbouring towns. There are also many urban and peri-urban sectors that are not connected to the EMAPICA network. There, the Juntas Administradoras de Servicios de Saneamiento and other community-based organizations provide drinking water to families, mainly from underground sources.

The current system for drinking water supply is highly unequal. On the one hand, there is a huge gap in daily water supply among sectors inside the EMAPICA network. Some sectors, in particular commercial and upper-class neighbourhoods, have 12-hour service, while the periphery, mostly poor areas, may have water just a couple of hours a day. Water quality is also a problem: some reports show that sulphate and manganese levels exceed allowable limits in several sectors, and users constantly complain about water turbidity. On the other hand, there is a gap between the sectors connected to the EMAPICA network and the ones that rely on local organizations. Although the municipal network system is far from efficient, often the water provision by community-based organizations is far more problematic: they can barely ensure daily water supply and water quality standards to their users (Gonzales, 2017).

In addition, the agro-export companies' increasing demand for groundwater deeply impacts the drinking water supply system, since they compete over the same source: the aquifers. Companies have invested in drilling more wells and deepening the existing ones to fulfil their increasing demand for groundwater, while EMAPICA and the community-based organizations have no budget for improving their water infrastructure. Because of aquifer overexploitation, several wells that were used to supply drinking water now have become unusable, tightening the water supply. Moreover, the agro-export boom has stimulated population growth and urban development, such as new housing projects, while EMAPICA has been unable to improve its service or extend its network due to its limited access to water and financial resources.

And in spite of their responsibility for generating water scarcity in the city, the agro-export companies and their organizations have had no role in searching for a solution. They have been very active in applying political pressure and even offering financial support to the regional government to facilitate new infrastructure projects that would increase the water availability on their productive land, but have had no interest in dealing with water scarcity in the city of Ica (Gonzales, 2017).

In 2013, the regional government designed a project to improve the EMAPICA supply system. The project anticipated a new water provision of 316 L/s, almost doubling the capacity of the system, to be distributed mainly among the 17 interconnected sectors (Gonzales, 2017). But the project has been opposed by the agriculture producers in Los Molinos, who say that their agricultural and domestic water supply would be severely affected. Despite the opposition, the project started in 2017. However, the new water supply will be used mainly to enhance daily water provision in middle-class residential neighbourhoods and commercial urban sectors, adding just a few new sectors to the network; most poor and peri-urban users will not see an improvement in the quantity or quality of their drinking water supply.

The Ica case shows us how public policies have prioritized infrastructure, institutions, and the use of water resources for agro-export users, to the detriment of the poor urban population and smallholders, materializing the hydro-territorial imaginary of the elites. So far, this reproduction of visible inequality in water services has not triggered major political unrest in the city of Ica. The agribusiness discourse about employment opportunities seems to be powerful enough to convince poor city dwellers that companies have the right to overexploit the aquifers – even to the detriment of more equal drinking water services. In Ica, public and private investments to improve water access have privileged agro-export rural zones over poor urban zones. In the emergent hydrosocial territory, the centre of power is not located in the urban centre but in the agro-export zones to which water flows.

Governing the hydrosocial territory: agribusiness's accumulation of political power

The redefinition of Ica's hydrosocial territory has resulted in a new political dynamic, with new territorial governance consolidating agribusiness elites' territorial dominance. Through groundwater boards they have dominant political and economic influence in the zone. Through these boards, or directly, the agro-export group has infiltrated various levels of government to keep their use of groundwater from being regulated and ensure their privileges of non-restricted exploitation of aquifers and over-extraction from wells (Damonte, 2015; Damonte & Gonzales, 2018; Urteaga, 2014). Here we mention some examples.

Similar to how governmental investments in urban-oriented water infrastructure are mainly directed towards the wealthy sectors (in which agribusiness representatives and other elite sectors dwell), the central and regional administration takes charge of infrastructure projects that almost exclusively benefit the agro-export companies of the Ica Valley. The government organization (PETACC) in charge of the canal that has carried water from the highlands of Huancavelica to the Ica River since the 1950s has now proposed to extend the canal that captures the water. This extra water would benefit the further expansion of the agro-export companies in the Ica Valley. This extension canal, called Ingahuasi, is fiercely opposed by the local highland communities of Pilpichaca and Carhuancho, because it will negatively affect the natural and human-created meadows they use to raise alpacas (Domínguez, 2014; Hoogesteger & Verzijl, 2015; Urteaga, 2014). Another infrastructure project proposed to strengthen their wished-for hydrosocial territory and provide more water for agribusiness is the diversion of water from the Pisco River southwards to the Villacuri pampa to recharge the aquifer.

Next, the best information on the territory's aquifers, wells, and volumes of water extracted is held by private producers. ANA has to get information from boards to attempt to inventory wells and calculate the volumes of water extracted. However, this information is relative, and the government has only approximate figures. Moreover, the limited information that the government does have is not made public without boards' consent, and regulation is based on producers' own declarations. As one public official put it, 'Agro-exporters tell us they know how to manage water; when we want to install flow meters, they tell us there is no point, because they have much more accurate meters already' (8/2016).

The users' board has records on wells and volumes, but the ANA has no way to verify them. Public officials have no access to agro-export companies and prefer to avoid confrontations with the owners of large farms. In the words of a former official, 'Agro-exporters have plenty of contacts in the government, political influence, and money, so if we get tough with them, they will get an order from our superiors to fire us' (7/2015).

Moreover, ANA estimates that informal wells have proliferated even during the water-ban period. About 65% of wells in Villacurí and 81% in Lanchas have no license at all (ANA, 2015). Clearly, the self-regulation policy is not working, but the lack of adequate government regulation seems to be politically motivated and key to the constitution of their territory, especially since small and medium well owners *are* regulated. The agro-exporters, in contrast, rely on their power and political legitimacy to avoid State oversight.

State institutions go along with the informal arrangements for access to groundwater that producers maintain and even legitimize in the context of consolidating the agro-export hydrosocial territory. Apart from land, water, and financial and technical resources, the territorial changes in Ica also exemplify how the rights and rules that used to protect small water users have now been amended to facilitate and protect investment of (international) capital and, consequently, accumulation of water use by large companies for export. Bans on extracting groundwater are the most important legal instrument to prevent over-pumping, but as was explained, in practice these laws are subtly adjusted or simply not enforced. In Ica, most new licenses – granted just before the official water ban – have been for agro-exporters. In the recent process of 'updating' water use rights in the Ica region, small users have been the ones who have lost rights, which the water authority has transferred to large farmers (Cárdenas, 2012).

The transnational agro-export discourse, promoted by companies and government alike, is strategic for legitimizing the new hydrosocial territory's imaginary and implementation.

Conclusions

This article has shown how hydrosocial territories have been redefined in the Ica Valley, as the result of the profound government-subsidized agro-export boom. In particular, it shows that the interlaced reconfiguration of different key components has resulted in the construction of an emergent agribusiness hydrosocial territory. Unlike other similar cases, this territorial construction has unfolded largely uncontested, under a hegemonic discourse of development. This article has analyzed the different territorial dimensions – physical, economic, institutional, political and discursive – that together produced the territorial redefinition.

In physical terms, the changes involve the expansion of the agricultural frontier by implementing capital-intensive infrastructure and technology to extract groundwater and produce new, highly water-consumptive luxury crops for the global North and China. In social terms, the changes reflect agricultural development led by agro-export companies making high-tech use of water. This change has brought new social relationships among traditional stakeholders in the watershed (small and medium owners) with emerging groups of agro-exporters and new contingents of wage-earning workers. At the same time, public investments in water are diverted from programmes that could sustain rural smallholders and the poorer segments of urban city population in Ica City.

The new social scenario is unfolding in a context dominated by the idea that water is a resource to be used primarily to generate economic capital. In political-administrative terms, the changes involve consolidating an agro-export project able to instrumentalize, commodify and discipline local and national societal rights and relationships, and penetrate throughout the new institutional structure for water management created by the State.

In Ica as in other Peruvian coastal valleys, the agro-export group's political and economic power has become the agent driving hydro-territorial redefinition. Supported by the last decades' government-promoted neoliberal policies and mindset, the agro-export companies establish their power by developing the financial and technical capacity to pursue high-tech export agriculture and contest existing territorialities and alternative imaginaries related to smallholders' livelihood, food-based agriculture, and dignified drinking water services for the cities. Next, they build on their political capacity to establish the rules of access to groundwater, overriding other organizations in the watershed including the State, and legitimize their water control using the dominant discourse. In this picture, private enterprise makes the desert bloom; it earns foreign exchange for the country and provides employment for the region; and it uses cutting-edge irrigation technology for 'efficient' water management, which must be followed to utilize other abandoned land in Ica and other valleys on the coast.

Instrumentally, agro-export and government agents define 'water scarcity' as if it were a phenomenon originating from nature's features and not from their own desires to (over)exploit the region's natural resources. The influence of agro-exporters and their transnational chains means that water bureaucracy, particularly local water authority offices, are unable to implement public institutional regulation procedures and must tacitly acquiesce to the informal institutions by which agro-exporters access groundwater, even during water emergencies in which the use of wells is banned.

The result of this hydro-political territorialization project is a capital-induced, monocultured and highly consumptive water extraction area in Peru's desert zone that produces not just luxury export crops for elites and remote continents but also localized urban water abundance and squandering, deep distributive inequality, legal-political injustice and environmental degeneration. This is not inevitable; there are less destructive viable alternatives.

Notes

1. See Boelens et al. (2016), Budds and Hinojosa (2012), Hommes and Boelens (2017, 2018), Hoogesteger et al. (2016), Swyngedouw (2009), and Vos and Hinojosa (2016).
2. http://www.ana.gob.pe/media/528,051/ica%20-%20plan%20de%20gesti%C3%B3n.pdf.
3. See also http://www.juasvi.com/juasvi_agro1.html.

Disclosure statement

No potential conflict of interest was reported by the authors.

ORCID

Rutgerd Boelens http://orcid.org/0000-0002-8412-3109

References

Allan, J. A. (2003). Virtual water – The water, food, and trade nexus. Useful concept or misleading metaphor? *Water International*, *28*(1), 106–113. doi:10.1080/02508060.2003.9724812

Autoridad Nacional del Agua. (2015). *Plan de Gestión del Acuífero de Ica-Villacurí-Lanchas*. Presentation (PPT). Lima, Peru: Autoridad Nacional del Agua.

Autoridad Nacional del Agua. (2018). Plan de Gestión del acuífero del valle de Ica y pampas de Villacurí y Lanchas. Retrieved from http://www.ana.gob.pe/media/528051/ica%20-%20plan%20de%20gesti%C3%B3n.pdf

Bakker, K. (2010). *Privatizing water. Governance failure and the world's urban water crisis*. Ithaca, NY: Cornell University Press.

Baviskar, A. (2007). *Waterscapes. The cultural politics of a natural resource*. Delhi: Permanent Black.

Boelens, R. (2014). Cultural politics and the hydrosocial cycle: Water, power and identity in the Andean highlands. *Geoforum*, *57*, 234–247. doi:10.1016/j.geoforum.2013.02.008

Boelens, R. (2015). *Water, power and identity. The cultural politics of water in the Andes*. London: Routledge.

Boelens, R., Hoogesteger, J., Swyngedouw, E., Vos, J., & Wester, P. (2016). Hydrosocial territories: A political ecology perspective. *Water International*, *41*(1), 1–14. doi:10.1080/02508060.2016.1134898

Boelens, R., & Vos, J. (2012). The danger of naturalizing water policy concepts: Water productivity and efficiency discourses from field irrigation to virtual water trade. *Agricultural Water Management, 108*, 16–26. doi:10.1016/j.agwat.2011.06.013

Bourdieu, P. (1998). *Acts of resistance against the tyranny of the market.* New York, NY: The New Press.

Budds, J. (2008). Whose Scarcity? The hydrosocial cycle and the changing waterscape of La Lingua River basin, Chile. In M. Goodman, M. Boykoff and K. Evered (Eds.), *Contentious geographies: Environmental knowledge, meaning, scale* (pp. 59–78). Aldershot, UK: Ashgate.

Budds, J., & Hinojosa, L. (2012). Restructuring and rescaling water governance in mining contexts: The co-production of waterscapes in Peru. *Water Alternatives, 5*(1), 119–137.

Cárdenas, A. (2012). *La carrera hacia el fondo. Acumulación de agua subterránea por empresas agro-exportadoras en el Valle de Ica, Perú.* Report. Netherlands: Wageningen University.

Chatterjee, P. (2007). *La Nación en Tiempo Heterogéneo y Otros Estudios Subalternos.* Lima: IEP, CLACSO, SEPHIS.

Corbridge, S., Wiliams, G., Srivastava, M., & Véron, R. (2005). *Seeing the state: Governance and governmentality in India.* Cambridge: Cambridge University Press.

Damonte, G. (2011). *Construyendo territorios: Narrativas territoriales aymaras contemporáneas.* Lima: GRADE & CLACSO.

Damonte, G. (2015). Redefiniendo territorios hidrosociales: Control hídrico en el valle de Ica, Perú (1993–2013). *Cuadernos de Desarrollo Rural, 12*(76), 109–133. doi:10.11144/Javeriana.cdr12-76.rthc

Damonte, G., & Gonzales, I. (2018). Política Pública, Relaciones de Poder y Conflicto de Intereses en la Gestión del Agua en Ica. In M. T. Oré & I. Muñoz (Eds.), *Aguas en Disputa: Ica y Huancavelica, entre el entrampamiento y el diálogo* (pp. 75–112). Lima: PUCP.

Damonte, G., Gonzales, I., & Lahud, J. (2016). La construcción del poder hídrico: Agroexportadores y escasez de agua subterránea en el valle de Ica y Villacurí. *Anthropologica, 34*(37), 87–114. doi:10.18800/anthropologica.201602.004

Damonte, G., Pacheco, E., & Grados, C. (2014). Dinámicas de Concentración y Escasez de Agua: El boom agro-exportador y los pequeños propietarios en la cuenca media y alta del río Ica. In M. T. Oré & G. Damonte (Eds.), *Escasez de agua en la cuenca del río Ica* (pp. 127–172). Lima: PUCP.

De Bont, C., Veldwisch, G. J., Komakech, H. C., & Vos, J. (2016). The fluid nature of water grabbing: The on-going contestation of water distribution between peasants and agribusinesses in Nduruma, Tanzania. *Agriculture and Human Values, 33*(3), 641–654.

Domínguez, I. (2014). Obras hidráulicas y aguas superficiales en la cuenca del río Ica, su valle y quebradas. In M. T. Oré & G. Damonte (Eds.), *Escasez de agua en la cuenca del río Ica* (pp. 55–86). Lima: PUCP.

Douglas, M. (1986). *How institutions think.* New York: Syracuse University Press.

Duarte-Abadía, B., & Boelens, R. (2016). Disputes over territorial boundaries and diverging valuation languages: The Santurban hydrosocial highlands territory in Colombia. *Water International, 41*(1), 15–36. doi:10.1080/02508060.2016.1117271

Duarte-Abadía, B., Boelens, R., & Roa-Avendaño, T. (2015). Hydropower, encroachment and the re-patterning of hydrosocial territory: The case of Hidrosogamoso in Colombia. *Human Organization, 74*(3), 243–254. doi:10.17730/0018-7259-74.3.243

Elden, S. (2010). Land, terrain, territory. *Progress in Human Geography, 34*(6), 799–817.

Foucault, M. (1991 [1978]). Governmentality. In G. Burchell, C. Gordon, & P. Miller (Eds.), *The Foucault effect: Studies in governmentality* (pp. 87–104). Chicago: University of Chicago Press.

Gonzales, I. (2017). *(Re) configuración de los territorios hidrosociales para la provisión de agua poblacional en la cuenca del Río Ica.* Paper for the Permanent Agrarian Research Seminar XVII, Cajamarca – Perú.

Harris, L. M., & Alatout, S. (2010). Negotiating hydro-scales, forging states: Comparison of the upper Tigris/Euphrates and Jordan River basins. *Political Geography, 29*, 148–156. doi:10.1016/j.polgeo.2010.02.012

Hendriks, J., & Boelens, R. (2016). Acumulación de derechos de agua en el Perú. *Anthropologica, 34*(37), 13–32. doi:10.18800/anthropologica.201602.001

Hommes, L., & Boelens, R. (2017). Urbanizing rural waters: Rural-urban water transfers and the reconfiguration of hydrosocial territories in Lima. *Political Geography, 57*, 71–80. doi:10.1016/j.polgeo.2016.12.002

Hommes, L., & Boelens, R. (2018). From natural flow to 'working river': Hydropower development, modernity and socio-territorial transformations in Lima's Rímac watershed. *Journal of Historical Geography, 62*, 85–95. doi:10.1016/j.jhg.2018.04.001

Hommes, L., Boelens, R., & Maat, H. (2016). Contested hydro-social territories and disputed water governance: Struggles and competing claims over the Ilisu Dam development in southeastern Turkey. *Geoforum, 71*, 9–20. doi:10.1016/j.geoforum.2016.02.015

Hommes, L., Boelens, R., Harris, L., & Veldwisch, G.J. (2019). Rural-urban water struggles: urbanizing hydrosocial territories and the evolving connections, discourses and identities. *Water International, 44*(2), 81–94. doi:10.1080/02508060.2019.1583311

Hoogesteger, J., Boelens, R., & Baud, M. (2016). Territorial pluralism: Water users' multi-scalar struggles against state ordering in Ecuador's highlands. *Water International, 41*(1), 91–106. doi:10.1080/02508060.2016.1130910

Hoogesteger, J., & Verzijl, A. (2015). Grassroots scalar politics: Insights from peasant water struggles in the Ecuadorian and Peruvian Andes. *Geoforum, 62*, 13–23. doi:10.1016/j.geoforum.2015.03.013

Ioris, A. (2016). Water scarcity and the exclusionary city: The struggle for water justice in Lima, Peru. *Water International, 41*(1), 125–139. doi:10.1080/02508060.2016.1124515

Joseph, G., & Nugent, D. (1994). Popular culture and state formation in revolutionary Mexico. In G. Joseph & D. Nugent (Eds.), *Everyday forms of state formation: Revolution and the negotiation of rule in modern Mexico* (pp. 3–23). Durham: Duke University Press.

Lefebvre, H. (1991). *The production of space*. Oxford: Blackwell.

Li, T. (2005). Beyond "the state" and failed schemes. *American Anthropologist, 107*(3), 383–394.

Linton, J., & Budds, J. (2014). The hydrosocial cycle: Defining and mobilizing a relational-dialectical approach to water. *Geoforum, 57*, 170–180. doi:10.1016/j.geoforum.2013.10.008

Long, N. (2004). Actors, interfaces and development intervention: Meanings, purposes and powers. In T. Kontinen (Ed.), *Development intervention. Actor and activity perspectives* (pp. 14–36). Helsinki: University of Helsinki.

Marshall, A. (2014). *Apropiarse del desierto: Agricultura Globalizada y Dinámicas Socioambientales en la costa peruana* (pp. 417). Lima: Instituto de Estudios Franceses.

Menga, F., & Swyngedouw, S. (2018). *Water, technology and the nation-state*. London: Routledge.

Migdal, J. S. (1988). *Strong societies and weak states: State-society relations and state capabilities in the third world*. Princeton: Princeton University Press.

Muñoz, I., Navas, S., & Milla, M. C. (2014). El problema de la disponibilidad de agua de riego. El caso de la cuenca del río Ica. In M. T. Oré & G. Damonte (Eds.), *Escasez de agua en la cuenca del río Ica* (pp. 87–126). Lima: PUCP.

Muñoz, I., & Zúñiga, A. (2018). El problema de sobreexplotación de los acuíferos de Ica y Villacurí-Lanchas: Estudio de caso bajo el enfoque institucional del manejo de los recursos de uso común. In M. T. Oré & I. Muñoz (Eds.), *Aguas en Disputa: Ica y Huancavelica, entre el entrampamiento y el diálogo* (pp. 112). Lima: PUCP.

Oré, M. T. (2006). *Agua Bien Común y Usos privados. Estado, riego y Conflictos en La Achirana del Inca*. Lima: PUCP.

Oré, M. T., Bayer, D., Chiong, J., & Rendón, E. (2012). La guerra por el agua en Ica, Perú. El colapso del agua subterránea. In E. Isch, R. Boelens, & F. Peña (Eds.), *Agua, Injusticia y Conflictos* (pp. 149-166). Lima: IEP, PUCP, Justicia Hídrica & CBC.

Oré, M. T., & Damonte, G. (Eds.). (2014). *Escasez de agua en la cuenca del río Ica*. Lima: PUCP.

Oré, M. T., & Muñoz, I. (Eds.). (2018). *Aguas en Disputa: Ica y Huancavelica, entre el entrampamiento y el diálogo*. Lima: PUCP.

Orlove, B., & Caton, S. (2010). Water sustainability: Anthropological approaches and prospects. *The Annual Review of Anthropology*, *39*, 401–415.

Progressio. (2010). *Drop by drop: Understanding the impacts of the UK's water footprint through a case study of peruvian asparagus*. London: Water Witness International.

Rodriguez-de-Francisco, J. C., & Boelens, R. (2016). PES hydrosocial territories: De-territorialization and re-patterning of water control arenas in the Andean highlands. *Water International*, *41*(1), 140–156. doi: 10.1080/02508060.2016.1129686

Seemann, M. (2016). Inclusive recognition politics and the struggle over hydrosocial territories in two Bolivian highland communities. *Water International*, *41*(1), 157–172. doi:10.1080/02508060.2016.1108384

Swyngedouw, E. (2009). The political economy and political ecology of the hydro-social cycle. *Journal of Contemporary Water Research & Education*, *142*, 56–60.

Swyngedouw, E., & Boelens, R. (2018). "... And not a single injustice remains": Hydro–Territorial colonization and techno-political transformations in Spain. In R. Boelens, T. Perreault, & J. Vos (Eds.), *Water justice* (pp. 115–133). Cambridge: Cambridge University Press. doi:10.1017/9781316831847.008

Urteaga, P. (2010). Ingenieria legal, acumulacion por desposesión y derechos colectivos en la gestion del agua. In R. Bustamante (Ed.), *Lo colectivo y el agua: Entre los derechos y las prácticas* (pp. 51–74). Lima: IEP.

Urteaga, P. (2014). Creadores de paisajes hídricos: Abundancia de agua, discursos y el mercado en las cuencas de Ica y Pampas. In M. T. Oré & G. Damonte (Eds.), *Escasez de agua en la cuenca del río Ica* (pp. 227–268). Lima: PUCP.

Verzijl, A., & Guerrero-Quispe, S. (2013). The system nobody sees: Irrigated wetland management and alpaca herding in the Peruvian Andes. *Mountain Research and Development*, *33*(3), 280–293. doi:10.1659/MRD-JOURNAL-D-12-00123.1

Vos, J., & Hinojosa, L. (2016). Virtual water trade and the contestation of hydrosocial territories. *Water International*, *41*(1), 37–53. doi:10.1080/02508060.2016.1107682

Wester, P., & Hoogesteger, J. (2011). Uso intensivo y despojo del agua subterránea. In R. Boelens, L. Cremers, & M. Zwarteveen (Eds.), *Justicia Hídrica: Acumulación, conflicto y acción social* (pp. 111–133). Lima: IEP.

Wittfogel, K. A. (1957). *Oriental despotism: A comparative study of total power*. New Haven: Yale University Press.

Payment for ecosystem services in Lima's watersheds: power and imaginaries in an urban-rural hydrosocial territory

Sonja Bleeker and Jeroen Vos

ABSTRACT

In Peru, payment for ecosystem services is an increasingly popular mechanism to secure the transfer of water from rural to urban areas. This article analyzes the process of setting up such a scheme in the watersheds of Lima. The concept of hydrosocial territories and a power analysis are used to scrutinize how urban-based imaginaries and top-down approaches result in a disregard of local knowledge, rationalities, history of urban–rural relations and land ownership structures in the highlands. This could result in unintended outcomes of the scheme and in subordinating upstream communities to the city's needs.

Introduction

By 2050, an estimated 66% of the world's population will live in urban areas (United Nations, 2014). This trend puts increased pressure on upstream water users and on the ecosystems of rural areas that are the hinterland that provides cities with water and other resources, such as food and energy. This raises important questions about rural–urban water dynamics and power struggles.

To guarantee continued quantity and quality of water supply to urban areas, an increasing number of drinking water companies and local governments have initiated conservation programmes in upstream catchment areas. These conservation programmes can entail direct territorial control (e.g., nature reserves or imposing land use restrictions) or voluntary programmes that provide incentives for conservation by the upstream land users. Since 2000, much attention has been paid to this latter form of conservation through payment for ecosystem services (PES) schemes (Schomers & Matzdorf, 2013). Most PES schemes employ governmental subsidies financed by taxpayers to improve water supply through ecosystem protection (Pigouvian-style payments). Others are based on Coasean-style market mechanisms, where providers of ecosystem services negotiate payments by the beneficiaries of those services.

A growing body of studies raises concerns about equity issues and additional limitations that PES schemes can place upon rural populations and rural water realities (Farley, Anderson, Bremer, & Harden, 2011; Pascual, Muradian, Rodríguez, & Duraiappah, 2010, Pascual et al., 2014; Rodríguez-de-Francisco & Boelens, 2014;

Rosa, Kandel, & Dimas, 2003). PES has drawn criticism from scholars characterizing PES initiatives as neoliberal endeavours converting community resources into commodities (Boelens, Hoogesteger, & Rodríguez de Francisco, 2014; Büscher, 2012; Fairhead, Leach, & Scoones, 2012; Fletcher & Büscher, 2017). Advocates regularly portray PES schemes as poverty-alleviation mechanisms because economic resources are paid to relatively poor rural dwellers (e.g., Pascual et al., 2010). However, powerful urban and rural actors are likely to profit most, as they are able to determine the rules of the schemes according to their interests (e.g., Corbrera, Kosoy and Tuna, 2007). In a context of unequal power relations between urban ecosystem services buyers and marginalized (and also internally unequal) rural communities as sellers, PES schemes are likely to have differentiating effects, especially for the poorest communities and community dwellers (Fletcher, 2010; Rodríguez-de-Francisco & Boelens, 2016). Furthermore, doubts have arisen about the environmental effectiveness of watershed-protection PES schemes, as hydrological effects are hard to attribute to specific measures (e.g., Calvet-Mir, Corbera, Martin, Fisher, & Gross-Camp, 2015).

In this article governmentality is used as an approach to study how dominant discourses and rationalities act as a form of power to shape people's subjectivities and behaviour. PES schemes are conceptualized as a water-based governmentality endeavour that establishes financial incentives to motivate upstream communities to behave and manage their territories according to an urban-watershed-conservation rationality. Governmentality refers to internalized beliefs about truth, values and norms that render the society governable (Dean, 2009). This will be further explained in the next section. The article details the results of a case study in the watersheds of Peru's capital city of Lima, where, besides existing and planned large-scale hydraulic infrastructure, a PES scheme is being set up as a new strategy for increasing the city's water security.

The case study demonstrates that PES promoters exercise different forms of power to shape Lima's urban-rural hydrosocial territories according to their ideas and interests. In other words, territorial imaginaries, discourses and participation strategies translate into certain land and water conservation principles that aim to guarantee urban water security. Even though reforms in water management legislation in Peru and other Andean countries increasingly incorporate the recognition of local water and land rights, history, knowledge and traditions, these reforms become instrumental to vested interests and are not operationalized to help marginalized rural communities (Roa-Garcia et al., 2015). In PES schemes the distribution of water and management of ecosystems are often naturalized, depoliticized and decontextualized (Zwarteveen & Boelens, 2014). The analysis of Lima's PES scheme shows that the shaping of hydrosocial territories is not only about the efficient management of natural resources but also about 'meanings, norms, knowledge, decision-making authority, and discourses' (Boelens, Hoogesteger, Swyngedouw, Vos, & Wester, 2016, p. 8).

This article is based on detailed ethnographic research carried out in Lima and the Rímac and upper Mantaro watersheds between March and September 2017. Special attention was paid to the rural community of Marcapomacocha, where 16 interviews with community authorities and members, a focus group with nine persons (community members and authorities, men and women), and observations in community assemblies were conducted. Seven interviews with community authorities of other communities were also held. In Lima, 19 representatives of public and private organizations were interviewed. To understand the development and changes of personal and institutional views over time, representatives of principal public organizations such as Lima's public water utility (SEDAPAL), the National Sanitation Services Regulator (SUNASS) that monitors the management of all Peruvian water utilities, and the National Water Authority (ANA) were interviewed more than once. Interviewees also included representatives of international and local NGOs that played an important role in promoting Lima's PES scheme. Finally, insights were gained from participation in different workshops and meetings, especially of the Natural Infrastructure Working Group,[1] which included the main public and private stakeholders of the PES scheme.

The article contributes to the ongoing discussion about PES principles and implementation that questions PES as an effective, democratic and equal way to safeguard socio-ecological systems. By focussing on a PES scheme in its initial set-up phase we illustrate how, from the start, the scheme is shaped by dominant imaginaries and interests, while being presented as apolitical and ahistorical. Specifically, the study of the Lima PES scheme shows the importance of considering land-tenure realities in rural communities involved in PES. The study is placed in the context of rural–urban dynamics, which is a theme of growing interest to scholars and practitioners in an urbanizing world with increasingly complex rural–urban relations.

First, the analytical framework of the article is briefly outlined. Thereafter, a description of Lima's watersheds and the rural communities involved in the scheme is presented to understand the context in which the PES scheme is introduced. Next, the process of setting up the scheme is examined, followed by the case study of Marcapomacocha. Finally, concluding remarks are presented, showing that unveiling the rationalities, imaginaries and power plays of the actors involved in the PES scheme allows a better understanding of how it transforms rural-urban hydrosocial territories through an ongoing process of negotiation and power plays.

Power plays shaping hydrosocial territories: conceptual notes

The conceptual approach of the article focuses on how hydrosocial territories are shaped through different forms of power, including governmentalities. This allows unravelling complex and interrelated dynamics shaped by visible, hidden and invisible power plays among stakeholders at various levels. Watershed-based PES schemes aim to conserve or restore ecosystems by reshaping and often commoditizing the relation between upstream 'water service providers' (often rural communities) and downstream urban or rural 'water-consuming' populations (Boelens et al., 2014). Because in most cases PES involves a diverse set of stakeholders with different and sometimes opposing spatial and political-geographical interests, knowledges and imaginaries, it needs to be considered as territorial politics and the outcome of negotiations and struggles between the involved actors (see Hommes, Boelens, Harris & Veldwisch, 2019, this volume).

Hydrosocial territories can be generally conceptualized as imagined and materialized 'spatial configurations of people, institutions, water flows, hydraulic technology and the biophysical environment that revolve around the control of water' (Boelens et al., 2016, p. 2). Stakeholders have divergent ideas about what constitutes a hydrosocial territory and its ideal state. Stakeholders' interests are best served if their view becomes the dominant view in decision-making spaces.

Gaventa's (2006) power cube is a framework for analyzing how power is used by actors across three dimensions that dynamically interact with and affect each other: forms, places, and spaces of power. The first dimension considers that power can be used in forms that are visible, hidden or invisible. Visible use of power takes place publicly. Hidden power is concealed from the general public: it is veiled, although in principle it can be seen. It often takes the form of 'agenda-setting' by powerful actors that control who and which topics are present at the decision-making table. Power is invisible when people internalize certain ideas and beliefs about their own powerlessness, incapacity or inferiority that make them think and act in a certain way. This third form of power equals the concept of governmentality, developed by Michel Foucault (1991), as governmentality entails not only the actions of governance but also the rationales and imaginaries that underlie those actions. Dean (2009, p. 18) defines governmentality as a form of power connected to 'any more or less calculated and rational activity, undertaken by a multiplicity of authorities and agencies, employing a variety of techniques and forms of knowledge, that seeks to shape conduct by working through the desires, aspirations, interests and beliefs of various actors, for definite but shifting ends and with a diverse set of relatively unpredictable consequences, effects and outcomes'. Studying the strategies employed by the most powerful to shape hydrosocial territories according to their interests and unravelling the power effects of different normalizing, invisible and discursive practices can reveal the political dimensions of these schemes. In Foucauldian terms, PES can be considered a governmentality project that entails different forms of power to 'conduct the conduct' of the less powerful (Foucault, 1991).

The second dimension of Gaventa's power cube is the places (or levels) where power is used and which can be global, national or local. Gaventa emphasizes the importance of connections or vertical links between these levels. In doing so, the power cube helps us analzse how PES (as a specific, dominant form of rural-urban hydrosocial territory) is shaped by decision making at local, national and global levels.

The third dimension is spaces, which are considered as arenas of engagement where different stakeholders have different power to make decisions. Spaces can be closed, invited or claimed. Closed spaces are situations where a limited group makes decisions. They can be bureaucrats, elected officials or experts, but they can also exist in private companies and civil society organizations. Invited spaces are situations where authorities invite certain stakeholders to take part in decision making. Most public consultations take place in such a space, where marginalized groups of people are usually not invited, as the agenda for discussion is set and only certain types of arguments are accepted. Claimed spaces are those created and conquered by organized groups of stakeholders themselves, using the media and sometimes international networks of like-minded people to ensure they can set the agenda and make decisions (Gaventa, 2006).

Gaventa's power cube, combined with the concepts of hydrosocial territory and governmentality, enables us to entwine a discursive with an institutional approach to power. This is particularly important when analyzing PES schemes, which are produced within a specific institutional setting and embedded in particular discourses. PES schemes are designed and operated according to institutional norms and rules that determine levels of participation, flows of finances and hierarchies in decision making. In this arena, both powerful and marginalized actors aim to operate on different levels of action, and open and close different spaces to foster their interests, using an ensemble of strategies that shape the outcomes of power struggles that are materialized in specific land and water management practices (Rodríguez-de-Francisco & Boelens, 2014).

Lima's watersheds

The fast-growing city of Lima is on the Pacific coast of Peru. It has almost 10 million inhabitants, or about one-third of the population of the entire country. Political power and over half of the country's economic activity are concentrated here. Lima's coastal zone is a desert, with average annual rainfall of only 9 mm. This leads to high dependency of the city on Andean watersheds as a main water source (Schütze & Robleto, 2010). However, environmental degradation in the headwaters of the rivers that provide water to Lima is destabilizing the hydrological cycle. It is caused by small-scale farming and cattle-raising, but also by mining and other legal and illegal commercial activities. These activities increase seasonal fluctuations in river flows, provoking more severe and prolonged water shortages in the dry season, and soil erosion, landslides and floods in the rainy season (Gammie & De Bievre, 2015).

Rural communities in the territories where the water for Lima originates are located in the upper parts of the three watersheds that surround Lima: Chillón, Rímac and Lurín (on the Pacific side of the Andean mountain range). Water is also transferred to Lima from the upper Mantaro watershed, which is on the Atlantic side (Figure 1). SEDAPAL and the hydro-electrical energy company ENEL (formerly EDEGEL) have installed large infrastructure projects to collect and transfer water from these parts of the Andean mountains to feed the Rímac River, which is the main water supply for Lima. The systems comprise dammed lagoons, channels, inter-basin tunnels and a trans-Andean tunnel that ultimately transfers water to the Rímac. During the dry season, the Marca systems accounts for about 60% of the total water supply for Lima (SEDAPAL, 2014).

Despite this hydraulic infrastructure, the Lima region faces a deficit of nearly 43 million cubic metres of water each year (Gammie & De Bievre, 2015; SEDAPAL, 2014). This deficit is addressed with groundwater. In a context of a growing urban population and increasing per capita use of water, SEDAPAL is in a race to augment the water supply by constructing additional reservoirs and making new water transfers from the headwaters of the Mantaro watershed (part of the Amazon basin) to the country's capital (SEDAPAL, 2014).

Although the already constructed massive hydraulic infrastructure is owned and controlled by SEDAPAL and ENEL, this has not directly led to rural communities being deprived of water. Nevertheless, the communities perceive 'injustices related to the unfair distribution of costs and economic benefits, loss of autonomy over their water resources and the socio-environmental impacts of territorial transformations' (Hommes & Boelens, 2017, p. 71). The mega-projects have thus created new dependencies and power relations and in consequence transformed rural-urban hydrosocial territories. In the community of Carampoma, a community member reflected: 'For ENEL we [the community] are not important. Our parents already negotiated and made agreements

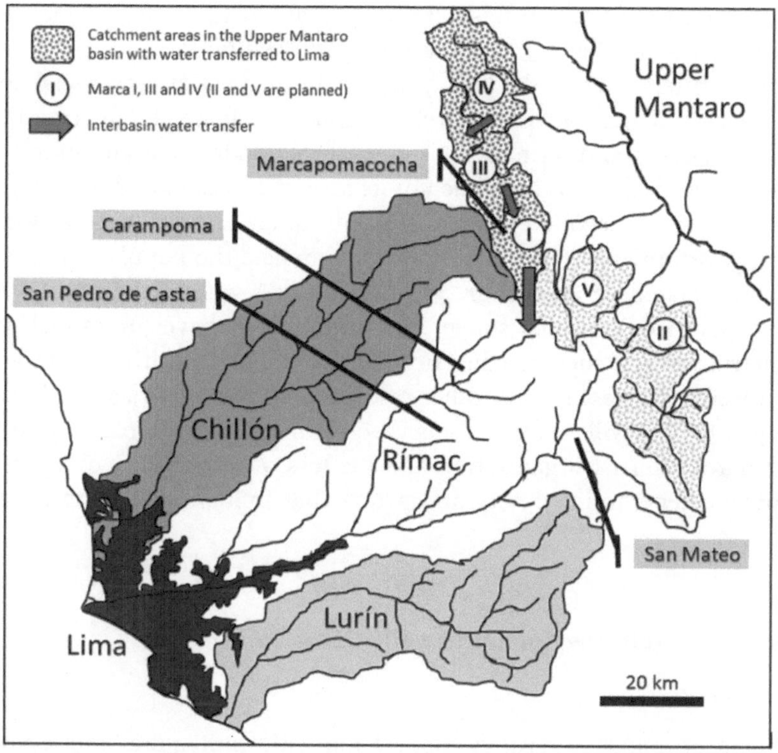

Figure 1. Interbasin water transfers and mentioned communities in Lima's watersheds.

[with the company] so we cannot change the situation a lot' (4 July 2017). In the community of Marcapomacocha, a community member commented, 'Nowadays water rights are officially arranged by the government. We have gone through a long process to get our licence. Now we have our share and the company has its share. But if Lima needs more water, SEDAPAL will get more water' (31 June 2017). These statements suggest that local communities took the historic inequalities between themselves and public and private companies (supported by governmental agencies such as the National Water Authority) for unchangeable and therefore unquestioned. It illustrates how communities have internalized the urban water users' and decision makers' right to impose legal rules and intervene in rural territories. This adoption of dominant ideology and forms of behaviour is a form of invisible power.

Since the 1960s, with the arrival of the hydroelectric company, water transfers, newly constructed access roads and development agencies, the communities have become increasingly physically and economically connected to Lima. In a case study of the rural community of San Pedro de Casta in the Rímac watershed, Ramírez (1980) found that the introduction of the community's economy to capitalism through improved physical access to urban markets and a Swiss development project for the increase and industrialization of milk production started a process of deconstruction of rural families' economy and disintegration of the community's organization and social systems. These processes of social fragmentation, weakening of internal organization and transformation of cultural identity because of increased rural–urban interaction are going on in many Andean communities (see also Boelens et al., 2012) and are specifically pronounced in Lima's watersheds (including the upper Mantaro).

Communities often refer to the lack of internal organization and social cohesion to explain their lack of capacity to effectively defend their interests, manage their socio-environmental systems and sustain their rural livelihoods. One community member said, 'Since the community has lost much of its former cohesion and identity, we do not have the capacity to … negotiate agreements with other actors. That's why it has been easy for local governments and companies to manipulate the population and introduce clientelism' (San Mateo de Huanchor, 31 June 2017). A young community member of Marcapomacocha made a similar statement: 'Although we have communal committees for many internal issues, such as education, environment and drinking water, they do not function well since people have other priorities. That is the tragedy of my community' (31 July 2017). As will be further analyzed for the case of Marcapomacocha below, the debilitation of internal organization and the loss of importance of agriculture as a principal income-generating activity are factors that influence the implementation of the PES scheme.

Setting up a PES scheme for Lima's drinking water

Historical context

The national government of Peru, in cooperation with multilateral and regional development agencies, has progressively reformed its water governance structures during the past decades. Since the 2009 Water Resources Law, water management has become increasingly organized according to Integrated Water Resources Management (IWRM) principles, with a multi-sectorial focus. Nevertheless, an uneasy coexistence exists between the simultaneous efforts of the state to, 'on the one hand, imbue water policy reform with international credibility via IWRM, and on the other, further economic development goals' (Mills-Novoa & Taboada, 2017, p. 370). For the latter purpose, the Peruvian government has traditionally implemented large-scale water infrastructure, with a clear focus on the supply side. Recently, PES schemes become a politically increasingly popular mechanism that complements hydraulic infrastructure to increase water supply, especially in the proximity of cities.

The Lima scheme is not the first PES scheme in Peru.[2] Since 2004, an alliance of public and private organizations (SUNASS and the Ministry of Environment [MINAM], together with the German International Cooperation (GIZ), the water utility of the city of Moyobamba and other local actors) implemented the first PES pilot in the Altomayo basin. The scheme was expected to directly benefit the inhabitants of Moyobamba and at the same time increase farmers' income (Bennett, Carroll, & Hamilton, 2013). However, quantitative data to support these expectations, as well as studies on the socio-economic impact on upstream communities, are scarce.

Based on the experience of Moyobamba, in 2012 Forest Trends and MINAM created the Incubator of Compensation Mechanisms for Ecosystem Services to advocate, and provide technical assistance for, the implementation of PES as a public policy, focussing specifically on Lima and other watersheds. The project actively involved public and private actors like ANA, SUNASS, SEDAPAL, Aquafondo

(Lima's water fund, launched by the Nature Conservancy and partners in 2010) and CONDESAN (Consortium for the Sustainable Development of the Andean Ecoregion, a regional NGO). These actors promoted water-related PES using multiple arguments and discourses. For example, they framed PES schemes as fitting into IWRM policies by establishing cooperation and dialogue between different water users (Andina, 2016; SUNASS, 2017c), as contributing to urban water security through secured watershed services (Gammie & De Bievre, 2015; SUNASS, 2017b), or as generating economic development for rural communities and thus creating a win-win situation (SEDAPAL, 2017).

As part of these efforts, SUNASS and the development agencies heavily (and successfully) lobbied to incorporate an article in the Modernization of Sanitation Law (Law 30045, 2013) that required Peru's 50 water companies to allocate a percentage of their water tariff to finance watershed conservation and restoration, and compensate upstream land and water users for better management practices. National and international policy makers and NGOs saw this law, together with the PES law (MINAM, Law 30215, 2016), as an innovative legal framework that would make Peru a forerunner of involving public water companies and other public and private actors in investment in ecosystem services in Latin America (Echavarría et al, 2015; Latin America Water Funds Partnership, 2017).

The creation of a PES scheme for Lima

In March 2015, SUNASS approved a resolution based on the Modernization of Sanitation Law that obliged SEDAPAL to invest 1% of its collected fees in watershed conservation or restoration projects. According to SEDAPAL, this will amount to about USD 30 million in five years (pers. comm., representative of SUNASS, 15 May 2017).

The active lobbying for and design of the PES scheme for Lima started several years before the 2015 resolution and provoked not only cooperation but also institutional competition between a wide range of actors and interest groups trying to position themselves in the 'ecosystem conservation and restoration' arena. For example, private associations such as Aquafondo and CONDESAN hoped to be contracted as project developer or implementer for SEDAPAL's funds, and had already started to seek funds from international donors for related activities. They competed to position themselves as organizations with the right expertise and knowledge (pers. comm., representatives of TNC and the French Institute of Andean studies, 25 March 2017). At the same time, public institutions showed a certain level of rivalry as well. A SUNASS representative said, 'The interest of ANA to be involved in the PES scheme relates to the fact that it is an opportunity to strengthen the river basin council [promoted by ANA]. However, river basin councils in general are ineffectual, and the potable water sector is not well represented. PES schemes are a more dynamic and effective way to create efficient and integrated water governance' (28 April 2017). The PES scheme was thus framed according to different institutional interests and triggered specific power dynamics, not only between rural and urban actors but also between different PES promoters in different regional and national spaces.

Hidden power in the design phase of the PES scheme

After the 2015 resolution that obliged SEDPAL to invest in ecosystem services, a lack of legal, technical and institutional tools and mechanisms, as well as SEDAPAL's initial disinterest in ecosystem conservation and restoration projects, prevented the scheme from being implemented until 2017. Initially, SEDAPAL showed resistance to the scheme since it did not perceive the PES scheme to be part of their competence as public water utility. Nevertheless, in compliance with the new legislation, PES money was collected and set aside in a special bank account. In this context, local and international NGOs involved in the process, in coordination with SUNASS and private consultants, elaborated studies that argued for and explained the benefits of the scheme for SEDAPAL, for example Gammie and De Bievre (2015), Aquafondo (2015), Kieser and Associates (2012), and CONDESAN (2016).

These studies aimed to motivate SEDAPAL to speed up the implementation of the scheme, providing evaluations of watershed conservation or restoration projects' potential to increase hydrological services from a cost-effectiveness and return-on-investment perspective. The proposed and evaluated projects were often referred to as 'green' or 'natural infrastructure', as opposed to 'grey infrastructure', which refers to conventional hydraulic works (cement dams, canals, tunnels, etc.). The main objective of the modelling studies was to demonstrate green infrastructure's economic competitiveness with conventional, 'grey' infrastructure solutions for water provision to downstream users. PES promoters presented the studies to both urban and rural water users as economic and technical instruments (for example in workshops and congresses), which gave them a certain legitimacy.

As a result of these studies, Aquafondo elaborated a portfolio of 33 project ideas in a total of 13 communities in the Rímac watershed, which were then included in SEDAPAL's five-year operational plan (2015–2020). Despite the studies' important role for decision making, the authors recognized that the outcomes presented high levels of uncertainty because of the lack of understanding of the impact of particular land uses on water flows. Even less did the studies take into account the complexities of land use as a human–environmental system 'that involves social-cultural, institutional and economic conditions together with geomorphology and soil characteristics' (de Lima, Krueger, & García-Marquez, 2017, p. 140). In this way, PES promoters shaped decision making on PES projects as a matter of identifying cost-effective ecosystem restoration and conservation interventions to increase the water supply. Interventions in complex human–ecological systems were decontextualized and depoliticized. This shows the exercise of hidden power (agenda-setting foregrounding positive effects and backgrounding uncertainties) and governmentalities that naturalize ideas about the relations between improved land use and improved hydrology, supposedly harmful land use practices of communities, the usefulness of economic cost–benefit analysis and the expected 'win–win' effects of PES schemes.

The start of SEDAPAL's PES scheme

With the designation of a new 'more ecosystem services minded' president of SEDAPAL, and under increased pressure from SUNASS, the utility started to take more leadership, and in June 2017 formulated their PES programme, called Sembramos Agua (roughly, Let's Sow Water). Based on previous studies and SEDAPAL's own hydrological and cartographic studies, the programme prioritized 15 micro-watersheds for the first interventions, mostly in the Rímac and upper Mantaro watersheds, complementary to the portfolio already included in the operational plan. The new Water and Sanitation Law (2016, legislative decree 1280) and the Directive on PES schemes of SUNASS (2017a) provided more clarity about the investment modalities of PES funds. Based on these norms, SEDAPAL considered two modalities for investing PES funds:

- Ecosystem conservation or restoration through the direct implementation of a public investment project by the water utility. The utility could decide to contract private organizations for this purpose. This modality considered community participation, but did not offer communities a direct incentive (monetary or non-monetary) for the provision of ecosystem services. The assumption was that communities would benefit economically because PES projects improve local water availability and quality, which enhances agricultural activities for income generation. This Pigouvian-style investment modality did not involve the creation of a market for suppliers and buyers of ecosystem services and was similar to other public watershed conservation efforts, which had had very mixed results in the past (Posthumus, Gardebroek, & Ruben, 2010).[3]
- Direct conditional payments to communities by the water utility, under the contractual obligation to implement certain land use practices to enhance the provision of hydrological ecosystem services. This Coasean-style investment modality is in line with the more traditional definition of PES schemes, offering direct monetary compensation to communities for better land and water management practices.

In the mire of technical studies, SEDAPAL's new leadership and debates between urban institutions and the communities' perspectives and interests were out of view, despite the emphasis both SUNASS and MINAM put on community involvement. For example, both SUNASS's directive on PES schemes and MINAM's PES law said that a PES scheme must involve establishing a platform for good governance, including communities. However, in Lima no such platform was implemented. ANA created a specific Natural Infrastructure Working Group as part of the Chillón, Rímac and Lurín River Basin Council. In practice, SEDAPAL led this group, which was in charge of formulating indicators for the PES programme and finally elaborating a Green Infrastructure Plan in 2017. This partly invited and partly closed space included all public and private organizations involved, except for rural representatives, and played an important role in the design of the PES scheme.

Another example of how communities' interests were not considered, despite the 'win-win' discourse, is the dominant focus on qualifying only the urban benefits from ecosystem services. Many of the planned projects had complementary elements, including training about watershed services and productive components such as installing sprinkler systems for irrigation of fodder crops (Aquafondo, 2015). The premise was that the boosted agricultural profits could compensate for the reduction in the number of animals or the exclusion of cattle from hydrologically important ecosystems such as wetlands, or even generate more profit than before. In the meetings of the Natural Infrastructure Working Group long discussions were held on the indicators that would be monitored to measure the scheme's success. As shown in the minutes of the meetings of the working group, of the 15 indicators identified, 14 focused exclusively on water availability (for example groundwater levels, number and discharge of local springs, base flow in rivers, etc.) and quality (erosion and soil loss, sediment in rivers, etc.), while other studies aimed at estimating the economic benefits for SEDAPAL (Gammie & De Bievre, 2015). One indicator referred to the number of strengthened rural organizations with awareness of the importance of ecosystem conservation. Although the group members actively promoted and assumed a pro-poor effect, socio-economic indicators of benefits for rural families were not included (GT-CAIN Cuenca CHIRILU, 2017).

The Natural Infrastructure Working Group was an example of an invited/closed space, in which 'power relations help to shape the boundaries of participatory spaces, what is possible within them, and who may enter, with which identities, discourses and interests' (Gaventa, 2006, p. 26). The group deployed hidden power, generated by the differences between, on the one hand, the rhetoric that promised empowerment and benefits for the marginalized, and on the other hand, the actual participation mechanisms provided by implementing agencies, which would involve people in predetermined projects to secure compliance with pre-defined goals (Cornwall, 2003).

While Lima's water-related PES will function at the watershed level, it was planned and negotiated at multiple scales involving local, national and global stakeholders. Rodríguez-de-Francisco and Boelens (2014, p. 359) described this process as the 'trickling down of global ideas into local projects'. International NGOs and development agencies (such as the US organizations Forest Trends and the Nature Conservancy, the Swiss Development Agency COSUDE, and GIZ) played an important role in influencing public policy via active engagement in the debate on PES and by commissioning studies that influenced decision makers by providing specific expertise and knowledge. As Gaventa (2006, p. 28) noted, 'the local, national and global are increasingly interrelated. Local forms and manifestations of power are constantly being shaped in relationship to global actors and forces, and in turn, local action affects and shapes global power.' Global players that promote PES schemes are empowered if they can present 'successful' cases. In that way, local activities empower global players, and global players shape national and local activities.

The case of Marcapomacocha

Lack of internal democracy and cultural erosion

Based on SEDAPAL's cartographic exercises, 10 of the 15 prioritized micro-watersheds were in the upper Mantaro watershed, three of them in the territory of the community of Marcapomacocha (Figure 1). Marcapomacocha has a territorial extension of approximately 15,000 hectares, of which almost all is natural pastures, wetlands and lagoons. These ecosystems are of importance for SEDAPAL since they naturally regulate the water flows that eventually are transferred to Lima. PES project ideas involved the improvement of pasture management by regulating grazing, which would increase infiltration of rainwater and reduce soil erosion.

In 2017 the community was home to about 800 persons. The community's altitude of over 4300 metres and corresponding climatic conditions limit the possibilities for agriculture, so sheep, cattle and camelid breeding provides about 60% of the community dwellers' income. Other economic activities include commercial activities and part-time construction work in cities, for mining companies or subcontractors of SEDAPAL that execute construction or maintenance works in the area (Provincia de Yauli, 2011).

A particularity of Marcapomacocha was that the communal assembly that made decisions relevant to the community, such as establishing the land use and land tenure system, consisted of only 50 persons, who were representatives of better-off families. The assembly functioned as a closed space that did not allow other families to become part of it. Although the land was communal property, in practice, the territory was divided between the 50 families that formed the assembly, and grazing plots were passed down from parents to children. This lack of internal democracy generated frustration among the other inhabitants of Marcapomacocha, even more so since 90% of the members of the assembly did not actually live in the village but resided and worked in Lima or in the nearby town of Huancayo. For them, income from livestock breeding was less important as primary economic activity. One elderly community member said, 'Families nowadays have livestock out of nostalgia but hire other people to herd the animals. Most of the time, they do not reside in Marcapomacocha' (31 June 2017). Livestock was considered an in-kind reserve of money that could be tapped in times of financial need. The people that did live permanently in the village rented their land from assembly members and were often hired by them to take care of the livestock.

Community members emphasized that the community's identity had eroded because of migration and economic diversification, but also because of inefficient and corrupt local governments over the last decades. According to them, people were easily influenced and manipulated and had a clientelist view of development, meaning they evaluated local politicians based on their promises and ability to provide their families with short-term gifts and economic benefits. In several interviews and the focus group discussion, community members mentioned that the erosion of social and communal structures was one of the causes of the degradation of local pastures and wetlands, since control over land use by communal authorities and sustainable traditional practices (such as rotational grazing) had decreased and land was semi-privatized. Conflicts with adjoining communities, as well as internal conflicts over land, were common.

Historical relations with SEDAPAL

Community members remembered the negotiations held with the hydropower company and SEDAPAL about the construction of the Marca system: 'They came and asked permission to use our land for the construction of the system, because there was a water emergency in Lima. With several big lagoons in our territory, how could we oppose that, even though of course we were worried about the impacts? (31 June 2017). Hommes and Boelens (2017, 2018) analyzed the discourses that SEDAPAL and the hydropower company used to push forward the water transfers and convince communities to cooperate. The authors show that the transfer project had been portrayed as inherent to modernity and justified with a discourse about the necessity of providing water for Lima. This made community members feel that it would have been morally unjust to oppose the transfer, but they still consider the monetary compensation they received very low. They blam both the company – 'they took advantage of our lack of knowledge and negotiation capacity' (2 July 2017) – and themselves, arguing that internal divisions and local corruption decreased their negotiating power. This reasoning about the morality of modernity, the rationale of water for the 'thirsty desert city' and the self-blaming because of internal division, are forms of invisible power in Gaventa's conceptualization, and examples of Foucauldian governmentalities. Marcapomacocha's community members stated that the construction of the Marca system's infrastructure caused floods and the erosion of several hectares of land, as well as the loss of many animals and even human lives due to long open canals crossing their territory. The community signed agreements with SEDAPAL for the remediation of these problems, such as the restoration of degraded areas and closing the open channels to prevent cattle falling in, but stated that the company had not complied with the arrangements. In July 2017 they were in the process of legally suing the company for large financial compensation. Even though many community members agreed that monetary compensation from SEDAPAL and other actors (hydropower and mining companies) in the past had not led to improved livelihoods, because of internal corruption and inefficient financial management, they emphasized the importance of economic benefits as a matter of equality. Hommes (2015, p. 44) notes that water transfers 'are also always associated with communities' hope of benefitting economically, be it through employment possibilities or monetary compensation. Accordingly, there was no case found in which a community was fundamentally against it, but instead, communities' concerns were initially connected to "getting a good deal".'

Setting up a PES scheme in Marcapomacocha

In September 2017, SEDAPAL started the first exploratory field visits and met with communal leaders of Marcapomacocha to explain the PES scheme, present the outcomes of technical studies to show the hydrologic and economic importance of PES and discuss project ideas with community leaders. While the hydraulic infrastructure was for managing water flows in the water bodies (lakes, streams and rivers), the PES scheme was meant to have a more direct impact by intervening in the management of the areas important for livestock raising, such as natural pastures and wetlands. The interventions were still under design, but options would entail the fencing of pastures and wetlands or improved management of camelids. Despite their conversations with SEDAPAL, the functioning and institutional arrangements of the PES scheme were not yet clear to the community. 'SEDAPAL planned to go to two other communities on the same day, so they did not have time to explain everything or to visit degraded areas as we [the community assembly] had planned. We know our territory and had ideas on where SEDAPAL could implement projects. Now we are not sure how exactly they plan to intervene. We need more information and explanation' (pers. comm., communal authority, 23 September 2017). This shows that the communities did not perceive SEDAPAL's field visits as effective for information exchange or joint decision making, and they felt that the ideas and knowledge they had were not taken into account. To interviewed community members it seemed that visits were just a pretence of inclusion, since they did not allow real community participation.

In practice, most families in Marcapomacocha (who worked for community members or rented land from them for livestock raising) were not represented in the communal assembly and lacked decision-making power over the territories that were vital for their families' economy. Those families were especially concerned about the potential impact of the PES-scheme projects on their livelihoods; they feared that they would have to abandon the areas prioritized for restoration by community members and SEDAPAL. 'What worries me is that we do not have any influence on this [PES] process, we do not even know what is discussed in communal assemblies' (pers. comm., non-assembly member of Marcapomacocha, 1 July 2017). Another person who rented land from assembly members said, 'I do not understand what PES is and what it implies for the area where my animals are grazing' (30 June 2017).

While talking about the PES scheme in the focus group, community members emphasized that the first thing they wanted is monetary compensation for the negative consequences of the hydraulic infrastructure. Some community members feared that the company were pretending to introduce the PES scheme in an attempt to settle the disputes over the negative impacts of the hydraulic infrastructure. 'We are not opposed to conserving our ecosystems; we will have to see what kind of projects they propose. But before talking about these types of projects, SEDAPAL has to formally reply to our complaints and provide the [financial] compensation we deserve' (pers. comm., communal advisor and former inhabitant of Marcapomacocha, 30 June 2017).

SEDAPAL characterized the claims of the community as exaggerated. 'Although I am also originally from an Andean community and I think we [SEDAPAL] have the obligation to be good neighbours to them and contribute to their development, I think that now they are only trying to take advantage of us. The damage they are referring to is minor, and SEDAPAL is already implementing projects to minimize further erosion of river banks.... Now they will also benefit from conservation [PES] projects' (pers. comm., representative of SEDAPAL, 30 June 2017).

The foregoing illustrates that SEDAPAL's imaginary of the catchment area for their water (their imaged hydrosocial territory) was one of bothersome communities that had always complained about the negative effects of infrastructure and demanded more compensation. The communal landowners had an imaginary of the watershed as their domain, where urban intruders came and damaged the landscape, while not recognizing the communities' problems and not paying sufficient compensation.

But despite the difficult relation with SEDAPAL, people were open towards the PES scheme, although they were sceptical about the local benefits of conservation and did not feel well informed about possible interventions and institutional arrangements. Most of the interest was generated by the direct financial compensation SEDAPAL planned. A water management specialist with experience in Lima's watersheds stated: 'Communities are very pragmatic. Whenever a third party presents itself and proposes the implementation of a specific development project, they rarely refuse, even though sometimes doubts exist about the positive impacts that these projects can have on their livelihoods. However, in a very pragmatic way, they just accept to see whatever benefit they may obtain, but without a real motivation to invest their own resources or time in the activities proposed' (pers. comm., rural development and water specialist, 4 June 2017).

In the focus group and interviews with community members and experts it became clear that although ecosystem conservation was seen as positive, concrete interest in investing time in such projects was very low. A former authority said, 'People are busy with their daily life and are not always in the community. Although they may want to engage in installing [green infrastructure] projects in exchange for payment, it is unlikely they will organize themselves to implement or maintain it, even with a financial incentive.' Another person from an adjoining village said, "Our children migrate and have another view of livestock-raising. They see it as a very uncertain activity in a harsh environment' (community member of San Mateo de Huanchor, 31 June 2017). The head of the Rímac office of the governmental Rural Agricultural Production Development Programme (AGRORURAL) made a similar observation: 'Our regulations require an in-kind financial contribution of communities in our projects, normally consisting of unskilled labour.... During the last decade, it has been increasingly difficult to motivate communities to contribute to forestation or terrace building. The pay they expect for labour is high, since they are used to receiving good salaries from mining and construction companies' (pers. comm., 12 June 2017).

These experiences and attitudes show that community members prioritized short-term economic benefits over the uncertain and long-term benefits of conservation of ecosystems that have lost importance for rural families' economies, especially for the better-off landowners, who rely more on urban-based activities for their income. Nevertheless, the risk of lack of interest or involvement of communities in PES projects was not a concern among PES promoters. They assumed that communities were intrinsically motivated to restore and conserve ecosystems and just needed some economic incentives and training to become natural ecosystem custodians. A member of the Natural Infrastructure Working Group said in one group meeting (16 October 2017), 'Communities will immediately grasp the benefits of green projects, because they understand the functions and working of nature in their territory.' As explained above, PES promoters focused on evidencing the benefits for SEDAPAL and downstream water users, while communities' benefits were not clearly defined.

Conclusions: power plays and the shaping of hydrosocial territories

Compared to other studies of PES schemes, this study highlights the importance of the design process in which the PES scheme was developed, as imagined by the PES promoters according to the interests of the water service utility. The study visualizes how in this initial phase dominant interests depoliticized the PES scheme and governmentalized and transformed imaginations about rural-urban hydrosocial territories through a system of top-down planning with many closed (or partially invited) spaces and forms of invisible power. Future follow-up research will have to reveal the execution of the projects and the material (changes in land use and water flows), economic (benefits and costs for all stakeholders) and political (power over the hydrosocial territory) effects in Lima's watersheds, and the actual acceptance of the scheme by the stakeholders.

The ideas behind the PES scheme were introduced to national policy makers and SEDAPAL by internationally and nationally operating NGOs. These PES promoters envisioned a hydrosocial territory governed by payments for ecosystem services provided by the upstream farmers and paid by the downstream urban water consumers. This image of the hydrosocial territory became the dominant vision in the national and local (watershed) decision-making place (in Gaventa's terminology).

PES promoters used hydrological and cost–benefit analyses to prove that PES was an effective and cost-efficient way to allocate SEDAPAL's financial resources. In this way, they reduced intervention in complex human–ecological systems to a technical and cost-effective way of increasing water supply. By making visible and calculable the relations between the hydrological functions of ecosystems on the one side, and urban water users' benefits on the other side, they made the complex relations that human–ecological systems entail depoliticized and governable. The framing of the PES scheme as rational and scientific is part of a complex power play aimed at creating seemingly neutral 'truths' that shaped rural actors' subjectivities and conducts.

While PES promoters focused on providing evidence of the positive return on investment of PES projects for urban water users, the same logic was not applied to communities. In PES promoters' imaginaries, rural communities were naturally motivated to participate in ecosystem restoration and conservation projects, since they would receive improved local ecosystem services and would at the same time benefit though economic compensation. The reality, however, is that communities did not consider PES or conservation projects particularly attractive and doubted their feasibility due to a weakened communal organization and historically complicated relations with SEDAPAL. This shows that in the proposed PES scheme the relations between humans and the natural world are perceived, valued and governed in line with dominant technical and economic imaginaries. The reasoning and imagining of desired behaviours are often at odds with upstream communities' ideas of compensation.

The use of invisible power (governmentality) was deployed through urban stakeholders' discourses about the absolute need of the city for water and communities believing the city has the right to control the water in the watershed. Communities seem to have adopted these 'truths' and also blamed themselves for their lack of negotiation capacity and their economic inequality because of internal division.

The study also highlights the PES promoters' disregard of the complexity of landownership in Marcapomacocha. The community dwellers who managed the grazing were not the owners of the land, owned only some of the animals, and feared increasing marginalization as a result of the PES scheme. This will most probably complicate the implementation of the PES projects. This specific complexity is not highlighted in other critical studies of PES schemes such as those of Rodríguez-de-Francisco and Boelens (2014), Büscher (2012), and Joslin and Jepson (2018).

Linking the concepts of hydrosocial territories, Gaventa's power cube and governmentalities allowed an unveiling of the rationales, imaginaries and power plays in the design of Lima's PES scheme. An economic, technical or institutional analytical framework would not have revealed the consciously and unconsciously deployed discourses that shaped the imagined hydrosocial territory in favour of the PES schemes, and how these linked to the institutional power (the closed spaces and multi-scalar places) of the PES promoters. The study illustrates that different hydrosocial territories exist, based on different knowledges and imaginaries, and that by studying the process of formation of, and struggle between, those divergent imagined hydrosocial territories it comes clear that certain hydrosocial territories become dominant by strategic application of closed spaces and multi-scalar influences (Boelens et al., 2016).

Given the increasing complexity of rural–urban relations, not taking into account rural inequities, rationalities and social, economic and historical contexts could render the PES scheme ineffective as a solution to urban water security while further subordinating upstream communities to the city's needs.

Notes

1. Grupo de Trabajo de Conservación e Infraestructura Natural de las Cuencas Chillón – Rímac – Lurín.
2. In Peru, PES schemes are now called compensation mechanisms for ecosystem services (*mecanismos de retribución por servicios ecosistémicos*). The renaming of the PES-like regulations is a sign of the uneasiness of many experts and policy makers with the market-based idea behind the PES principle. Despite this, the compensation mechanisms in Peru do not entail a fundamentally different idea.
3. Such as the former National Project for the Management of Hydrographic Basins and Soil Conservation (PRONAMACHS), which is now AGRORURAL, part of the Ministry of Agriculture.

Disclosure statement

No potential conflict of interest was reported by the authors.

References

Andina. (2016). Sunass y Autoridad Nacional del Agua cautelarán conservación de los recursos hídricos. Retrieved from http://andina.pe/agencia/noticia.aspx?id=619636

Aquafondo. (2015). Portafolio de Proyectos AQUAFONDO y SEDAPAL. Ocho tipos de intervenciones en infraestructura verde en la cuenca alta y media del río Rímac.

Bennett, G., Carroll, N., & Hamilton, K. (2013). Charting New Waters. State of Watershed Payments 2012.

Boelens, R., Duarte, B., Manosalvas, R., Mena, P., Avendaño, T. R., & Vera, J. (2012). Contested territories: Water rights and the struggles over indigenous livelihoods. *International Indigenous Policy Journal*, *3*(3), 5. doi:10.18584/iipj.2012.3.3.5

Boelens, R., Hoogesteger, J., & Rodríguez de Francisco, J. C. (2014). Commoditizing water territories: The clash between Andean water rights cultures and payment for environmental services policies. *Capitalism Nature Socialism*, *25*(3), 84–102. doi:10.1080/10455752.2013.876867

Boelens, R., Hoogesteger, J., Swyngedouw, E., Vos, J., & Wester, P. (2016). Hydrosocial territories: A political ecology perspective. *Water International*, *41*(1), 1–14. doi:10.1080/02508060.2016.1134898

Büscher, B. (2012). Payments for ecosystem services as neoliberal conservation: (Reinterpreting) Evidence from the Maloti-Drakensberg, South Africa. *Conservation and Society*, *10*(1), 29–41. doi:10.4103/0972-4923.92190

Calvet-Mir, L., Corbera, E., Martin, A., Fisher, J., & Gross-Camp, N. (2015). Payments for ecosystem services in the tropics: A closer look at effectiveness and equity. *Current Opinion in Environmental Sustainability*, *14*, 150–162. doi:10.1016/j.cosust.2015.06.001

CONDESAN. (2016). Análisis de los servicios ecosistémicos hídricos del sistema Marcapomacocha en la cuenca alta del Mantaro dentro del marco del MRSE de Lima. (3-35).

Corbera, E., Kosoy, N., & Tuna, M. M. (2007). Equity implications of marketing ecosystem services in protected areas and rural communities: case studies from meso-america. *Global Environmental Change*, *17*(3–4), 365–380. doi:10.1016/j.gloenvcha.2006.12.005

Cornwall, A. (2003). Whose voices? Whose choices? Reflections on gender and participatory development. *World Development*, *31*(8), 1325–1342. doi:10.1016/S0305-750X(03)00086-X

de Lima, L. S., Krueger, T., & García-Marquez, J. (2017). Uncertainties in demonstrating environmental benefits of payments for ecosystem services. *Ecosystem Services*, *27*, 139–149. doi:10.1016/j.ecoser.2017.09.005

Dean, J. (2009). *Democracy and other neoliberal fantasies: Communicative capitalism and left politics*. Durham, NC: Duke University Press.

Echavarria, M., Zavala, P., Coronel, L., Montalvo, T., & Aguirre, L. (2015). Green Infrastructure in the Drinking Water Sector in Latin America and the Caribbean: Trends, Challenges, and Opportunities. EcoDecisión.

Fairhead, J., Leach, M., & Scoones, I. (2012). Green grabbing: A new appropriation of nature? *Journal of Peasant Studies*, *39*(2), 237–261. doi:10.1080/03066150.2012.671770

Farley, K., Anderson, W., Bremer, L., & Harden, C. (2011). Compensation for ecosystem services: An evaluation of efforts to achieve conservation and development in Ecuadorian paramo grasslands. *Environmental Conservation*, *38*, 393–405. doi:10.1017/S037689291100049X

Fletcher, R. (2010). Neoliberal environmentality: Towards a poststructuralist political ecology of the conservation debate. *Conservation and Society*, *8*(3), 171. doi:10.4103/0972-4923.73806

Fletcher, R., & Büscher, B. (2017). The PES conceit: Revisiting the relationship between payments for environmental services and neoliberal conservation. *Ecological Economics*, *132*, 224–231. doi:10.1016/j.ecolecon.2016.11.002

Foucault, M. (1991). *The Foucault effect: Studies in governmentality*. Chicago: University of Chicago Press.

Gammie, G., & De Bievre, B. (2015). Assessing green interventions for the water supply of Lima, Peru. Cost-Effectiveness, Potential Impact, and Priority Research Areas.

Gaventa, J. (2006). Finding the spaces for change: A power analysis. *IDS Bulletin*, *37*(6), 23–33.

Gt-Cain Cuenca Chirilu. (2017) (Grupo de Trabajo "Conservación del Agua e Infraestructura Natural - Cuencas Chillón-Rímac-Lurín). Ayuda memoria del taller de lineamientos para la elaboracion de la linea base de las cuencas Chillón-Rímac-Lurín".

Hommes, H., Boelens, R., Harris, L., & Veldwisch, G.J. (2019). Rural-urban water struggles: urbanizing hydrosocial territories and evolving connections, discourses and identities. *Water International*, *44*(2), 81–94. doi:10.1080/02508060.2019.1583311

Hommes, L. (2015). *Urbanized waters, rural-urban linkages and hydro-social territories in Lima's watersheds* (MSc. Thesis). Water Resources Management Group. Wageningen University.

Hommes, L., & Boelens, R. (2017). Urbanizing rural waters: Rural-urban water transfers and the reconfiguration of hydrosocial territories in Lima. *Political Geography*, *57*, 71–80. doi:10.1016/j.polgeo.2016.12.002

Hommes, L., & Boelens, R. (2018). From natural flow to 'working river': Hydropower development, modernity and socio-territorial transformations in Lima's Rímac watershed. *Journal of Historical Geography*, *62*, 85–95. doi:10.1016/j.jhg.2018.04.001

Joslin, A. J., & Jepson, W. E. (2018). Territory and authority of water fund payments for ecosystem services in Ecuador's Andes. *Geoforum*, *91*, 10–20. doi:10.1016/j.geoforum.2018.02.016

Kieser & Associates. (2012). Identification of common project goals and a common metric for the Lima waterfund. Technical Report.

Latin America Water Funds Partnership. (2017). A proactive water management approach in Peru. Natural infrastructure as a key element in public policy. Retrieved from http://waterfunds.org/esp/wp-content/uploads/2017/05/Paper-Peru-A-proactive-water-management-approach-in-Peru-082018.pdf

Mills-Novoa, M., & Taboada, R. (2017). Coexistence and conflict: IWRM and large-scale water infrastructure development in Piura, Peru. *Water Alternatives*, *10*(2), 370–394.

Pascual, U., Muradian, R., Rodríguez, L. C., & Duraiappah, A. (2010). Exploring the links between equity and efficiency in payments for environmental services: A conceptual approach. *Ecological Economics*, *69*(6), 1237–1244. doi:10.1016/j.ecolecon.2009.11.004

Pascual, U., Phelps, J., Garmendia, E., Brown, K., Corbera, E., Martin, A., ... Muradian, R. (2014). Social equity matters in payments for ecosystem services. *BioScience*, *64*(11), 1027–1036. doi:10.1093/biosci/biu146

Posthumus, H., Gardebroek, C., & Ruben, R. (2010). From participation to adoption: Comparing the effectiveness of soil conservation programs in the Peruvian Andes. *Land Economics*, *86*(4), 645–667. doi:10.3368/le.86.4.645

Provincia de Yauli. (2011). Plan de Desarrollo Concertado de la Provincia de Yauli - Junín 2012–2021.

Ramirez Villacorta, Y. (1980). La penetración capitalista en una comunidad Campesina: El Caso de San Pedro de Casta, Huarochiri. *Debates en Sociología*, *05*, 39–70.

Roa-García, M. C., Urteaga-Crovetto, P., & Bustamante-Zenteno, R. (2015). Water laws in the Andes: A promising precedent for challenging neoliberalism. *Geoforum*, *64*, 270–280. doi:10.1016/j.geoforum.2013.12.002

Rodríguez-de-Francisco, J. C., & Boelens, R. (2014). Payment for environmental services and power in the Chamachán watershed, Ecuador. *Human Organization*, *73*(4), 351–362. doi:10.17730/humo.73.4.b680w75u27527061

Rodríguez-de-Francisco, J. C., & Boelens, R. (2016). PES hydrosocial territories: De-territorialization and re-patterning of water control arenas in the Andean highlands. *Water International*, *41*(1), 140–156. doi:10.1080/02508060.2016.1129686

Rosa, H., Kandel, S., & Dimas, L., 2003. Compensation for environmental services and rural communities. Programa salvadoreño de investigación sobre desarrollo y medio ambiente, San Salvador.

Schomers, S., & Matzdorf, B. (2013). Payments for ecosystem services: A review and comparison of developing and industrialized countries. *Ecosystem Services*, *6*, 16–30. doi:10.1016/j.ecoser.2013.01.002

Schütze, M., & Robleto, G. (2010). Challenges of water and wastewater management in the desert megacity of Lima/Peru-how can macromodelling help? Novatech.

SEDAPAL. (2014). Plan Maestro de los Sistemas de Agua Potable y Alcantarillado (Lima).

SEDAPAL. (2017). Poblaciones rurales pueden generar empleo como "emprendedores del agua". Nota de Prens N° 69. Retrieved from http://www.sedapal.com.pe

SUNASS. (2017a). Directiva de mecanismos de retribución por servicios ecosistémicos hídricos - MRSE hídricos. Resolución de Consejo Directivo Nº 045-2017-SUNASS-CD. Retrieved from: http://www.sunass.gob.pe/doc/normas%20legales/2017/re45_2017cd.pdf

SUNASS. (2017b). SUNASS impulsa la formulación del Plan Maestro de Infraestructura Verde de Sedapal. Retrieved from http://www.sunass.gob.pe/websunass/index.php/noticias/item/1043-sunass-impulsa-la-formulacion-del-plan-maestro-de-infraestructura-verde-de-sedapal

SUNASS. (2017c). SUNASS promoverá participación de EPS en la gestión integrada de los recursos hídricos a través de Convenio con MINAM. Retrieved from http://www.sunass.gob.pe/websunass/index.php/noticias/item/1152-sunass-promovera-participacion-eps-gestion-integrada-recursos-hidricos-convenio-minam

United Nations. (2014). Department of Economic and Social Affairs, Population Division (2015). World Urbanization Prospects: The 2014 Revision, (ST/ESA/SER.A/366).

Zwarteveen, M. Z., & Boelens, R. (2014). Defining, researching and struggling for water justice: Some conceptual building blocks for research and action. *Water International*, *39*(2), 143–158. doi:10.1080/02508060.2014.891168

OPEN ACCESS

CONCLUSIONS

Evolving connections, discourses and identities in rural–urban water struggles

Lena Hommes, Gert Jan Veldwisch, Leila M. Harris and Rutgerd Boelens

Introduction

In this special issue, we set out to analyze the dynamics, discourses, identities and material conditions that evolve with increasing urbanization and associated transformations of rural–urban (dis)connections. Considering the growing importance of cities, water is a particularly useful lens through which to understand the ways in which rural–urban territories and flows - consisting of interlaced material, social and symbolic components - are dynamically and continuously reconfigured. Focusing on water connections and disconnections that link rural and urban spaces necessarily involves attention to struggles over water use and control as well as over decision making, identities, discourses and associated politics of recognition. As many of our case studies highlight, new inequities and injustices are a key element of these changing social and material flows, connections and networks. In particular, the articles in this issue have explored the ways urbanization and rural–urban transformations reconfigure territories and power relations, with varying consequences for different water users, and the changing scales and pathways of water flows through these interlinked systems, as well as the broader hydrosocial dynamics at play.

Our introduction (Hommes, Boelens, Harris, & Veldwisch, 2019) outlined several points of departure for the case analyses in this issue. First, we described how urbanization needs to be understood as a process of constant making and remaking of multi-scalar territorial networks and interactions that extend beyond traditional city boundaries. As such, these processes reconfigure rural–urban hydrosocial territories and interlinkages physically, politically, technologically, economically and culturally. Second, the fact that urbanization is not a territorially discrete and self-contained phenomenon draws attention to the understanding that the sometimes neatly presented rural–urban dichotomy is misleading, as administratively or arbitrarily set boundaries are constantly transcended by movements of water, people, goods and ideas. Third, the connections and dynamics resulting from urbanization and cities' quest for secure supplies of energy, food, water and other resources produces divergent effects for different water users. This makes it important to inquire how exactly, by whom and in what ways inequalities in water access and control, representation and inclusion, and cultural-political recognition among rural and urban subjects are produced, reconfigured and materialized in new territorial reconfigurations and relationships. By bringing these contributions together in a single special issue, we are able to see multiple pathways of these changes, but also key shared elements between them. In all the cases, there is a complex retracing and reconfiguration of rural–urban geographies. In each case, focus on water flows, technologies and justice considerations provides a revealing and fruitful lens to think about the consequences and significance of these reconfigurations – for diverse locales, populations and ecologies.

The contributions to this issue provide numerous examples of how water embodies, transcends, and rematerializes important rural / peri-urban / urban connections (and sometimes disconnections). Traditional transfers of water through sturdy material structures from rural to urban areas are one example; payment for ecosystem services (PES), currently hyped as a win-win solution for both rural and urban populations, is another (see e.g. the Peruvian case analysis by Bleeker & Vos, 2019). Likewise, virtual water contained in food crops, and hydropower produced for growing urban populations, link people and territories along the rural–urban continuum through water. What all these old and new approaches have in common is that they rely on a myriad of techniques: they are produced by, and embedded in, specific institutional and legal arrangements, actor alliances, discourses, interests and technologies. With this multitude of actors and techniques that aim to secure water for urban areas and that thereby automatically reconfigure rural–urban hydrosocial territories, clashes between interests, projects or ideas are common and can lead to contestations and struggles, both fierce and subtle, but also some cases, to acceptance, compromise and adjustment.

Below, we elaborate on how the contributions of this issue exemplify and enrich our understanding of the above aspects of a political ecology of rural–urban water struggles. The contributions are diverse geographically as well as thematically and conceptually. Here we draw together the insights that they allow us to gain concerning new trends and issues. Finally, we explore a future research agenda that sets the base for working towards democratic rural / peri-urban / urban relations, where water users engage with each other on just terms.

Evolving rural–urban connections and technologies

As outlined, water supply projects connect far-away rural territories with the city, impacting rural environments, livelihoods, political positioning and identities (Hommes et al., 2019, this issue). Similarly, powerful agro-export companies delivering to urban populations around the globe divert water to the detriment of nearby towns (see the case in this issue from Peru: Damonte & Boelens, 2019; see also Hommes, Boelens, Duarte-Abadía, Hidalgo, & Hoogesteger, 2018; Veldwisch, Franco, & Mehta, 2018). These findings emphasize that water supply infrastructure is an example of creating both boundaries and connections in which water flows link places, people and practices and thus shape hydrosocial territories – a dynamic reconfiguration that is often deeply conflictive (Boelens, Hoogesteger, Swyngedouw, Vos, & Wester, 2016).

The articles in this issue demonstrate that claims about urban water needs and the indispensability of supply projects (or flood protection, in some cases) often have both naturalizing and universalizing components (Roth, Zwarteveen, Joy, & Kulkarni, 2018; cf. Bakker, 2010; Lynch, 2013; Mehta, 2006). First, naturalizing, in cases where lack of access to drinking water in cities is portrayed as a problem arising because of environmental conditions such as drought – carefully avoiding questions about the socio-political aspects that have caused some people to lack access while others have all they need (in this issue, see Damonte & Boelens, 2019; Goldman & Narayan, 2019; cf. Harris, Goldin, & Sneddon, 2013). Another example is the political decision to inundate rural areas to protect cities from flood risks that are presented as a natural problem or as a result of climate change (in this issue, see Hidalgo-Bastidas & Boelens, 2019; cf. Porter & Demeritt, 2012; Warner, 2010). Second, universalizing, when lack of access to drinking water and sanitation are portrayed as a universal problem of today's urban world, rather than socially and politically produced (in this issue, Torio, Harris, & Angeles, 2019; cf. Aguilera-Klink, Pérez-Moriana, & Sánchez-García, 2000). This universalizing logic also leads to the neglect of history and context-specific processes that require locally particular or tailor-made solutions rather than one-size-fits-all water policies that aim to solve urban water needs through universalistic water supply approaches. While the lack of access is certainly a problem that needs to be tackled in many places, such references to a supposed universalism (which is implicit even in the Human Right to Water; see Harris, Rodina, & Morinville, 2015) may overlook the fact that locally experienced lack of access is often an outcome of decades-old histories of complex socio-environmental and political processes that affect populations differentially. Instead, universalizing notions paper over inequalities within and between urban and rural areas, and the socio-political reasons for unequal positions within these heterogeneous spaces. The contributions to this issue have made these dynamics explicit and a focal point for analysis, rather than glossing over the context specificities, histories, and associated power dynamics.

This points to another crucial issue when discussing policies, rural–urban water struggles and newly evolving connections, discourses and identities: it bears repeating that the rural and the urban are not clear-cut, distinguishable, nor homogeneous spaces. Rural–urban studies have made clear how rural and urban domains strongly overlap and increasingly entwine. In this issue, for instance, Hoogendam (2019) accentuates the complex but crucial role of peri-urban spaces for rapidly changing water governance frameworks in Bolivia. These spaces with profoundly dynamic and porous borders have a particular importance, being situated in between the categories of 'urban' and 'rural' – in geographical terms as well as culturally, politically and normatively. Next, also within these categories, all rural and urban spaces contain a large diversity of actors, each with their particular demands, interests, identities, powers and, accordingly, access to and control over water resources. Goldman and Narayan (2019), for example, point out that 'urban' is not *one* homogeneous water-user group, but a contested space in itself, with differentiated access regimes between urban peripheries and privileged (business) areas. In a similar manner, Hoogendam's article shows how rural spaces in Cochabamba, Bolivia, harbour diverse actors and interests that are continuously changing over time, in the process reshaping hydro-territorial notions, water claims and water solutions. This clearly demonstrates the fluidity of spaces, people and water, and consequently calls for caution in analyzing rural–urban relations along and within clear-cut categories. Indeed, as various scholars have shown, the reality is much more complex, dynamic and interwoven: people, ideas and water flows are not fixed in only one particular category of space. Wessels, Veldwisch, Kujawa, & Delcarme (2019) illustrate that historic privilege can also align very differently with social groups along the rural–urban spectrum, with power centred in a white rural elite profiting from high-tech irrigated agro-export production while successfully portraying historically disadvantaged groups of black urban dwellers, who still lack access to basic sanitation and water supply services, as polluting their water sources. Here, an analytical lens that highlights urban water flows, discourses and connections helps make these equity concerns and power dynamics clear

As these Bolivian and South African cases manifest, power dynamics often include cultural politics and specific discursive practices in which identities and legal-political-cultural categorizations of water user groups are framed, assigned, deployed and also self-constructed: between 'the urban' and 'the rural' and also inside, to either legitimize or invalidate claims to particular water resources and services. The cases of Ecuador (Hidalgo-Bastidas & Boelens, 2019), Bangalore (Goldman & Narayan, 2019), Spain (Duarte-Abadía & Boelens, 2019) and Peru (Damonte and Boelens, 2019) also show this strategic-political (and sometimes unconscious) use of identity framing to justify rural–urban water transfer projects and undemocratic water decision making: it often relates to dualistic and stereotyped images of modernity (e.g., city) and backwardness (e.g., countryside); professional expertise (e.g., scientists and hydraulic experts) and non-formal knowledge (e.g., peasants); productivity (e.g., export growers) and wastefulness (subsistence farming) – but equally relates to class-, gender-, caste- and ethnicity-related labelling. As these same contributions show, cultural-political classifications and prioritization also convey water-related identity labels that speak of 'water consumers' (which makes seeing these subjects as water governance protagonists very difficult), 'clients' (assigning water users a specific role in the market), 'beneficiaries' (which delegitimizes any questioning of water-transfer projects), et cetera. In many cases, such labelling or assigned roles are inscribed in, and in turn consolidated by, specific infrastructure designs, the most prominent example being the functioning water tap in urban households that creates 'clients' in some neighbourhoods and 'non-clients' in other, often historically already disadvantaged zones (see in this issue Torio, Harris & Angeles, 2019; Wessels et al., 2019). In this way, water infrastructure can be both a source and the result of particular unequal social and political positions.

The articles also show the need to look beyond those conceptualizations of rural–urban hydrosocial territories that assume rather bounded geographies and fixed, limited spaces, in order to comprehend the broader rural–urban water interactions and struggles (Hoogesteger, Boelens, & Baud, 2016; Veldwisch et al., 2018; Vos & Hinojosa, 2016). Rural areas and cities are also deeply interwoven in global dynamics, through the flow of goods, ideas, money and people from places far beyond the nearby rural areas. Water philosophies and ideologies promulgated at the global level affect cities and their water management decisions. The examples are numerous, ranging from neoliberal economic policies to Integrated Water Resources Management paradigms that may originate in global arenas but play out in particular shapes in local spaces (Harris, 2015; Swyngedouw, 2015; Zwarteveen, 2015). Even though it is not the primary focus of this issue to show the impact of global dynamics on place-bound local rural–urban territories, global aspects are inevitably to be found in a number of studies. For example, Goldman and Narayan (2019) show how global capital and ideas about solutions for Bangalore's water scarcity affect the city's hydrosocial regime and the different water user groups. Likewise, cases from South Africa and Peru (Wessels et al. and Damonte & Boelens, 2019) show how foreign (largely urban) market demands for certain agricultural products are reconfiguring local rural–urban spaces and relations, promoting water flows to profitable large-scale agro-business (to be exported as product-embedded 'virtual water') rather than to marginalized urban users. PES implementation in Lima is a clear example of how globally promoted ideas materialize and become adopted in a local context, and in turn feed back into global policy discussions as implementation examples (Bleeker & Vos, 2019). Thus, even though research might be focused primarily on local dynamics, political ecology approaches at times make visible the linked multiscalar processes that entwine scales and processes from local to global (e.g., Harris, 2015; Hommes et al., 2018; Swyngedouw & Boelens, 2018).

Water struggles and social mobilization

The different contributions to this issue make the politics of rural–urban water flows explicit by engaging with some of the connected discussions and the existing diversity of contestations and mobilizations, analyzing different degrees, ways and actor alliances.

First, there are those cases where urban water supply projects are fiercely contested by communities in the areas from where water is to be diverted. As Hoogendam (2019) illustrates, in the case of Cochabamba an irrigators' movement formed to protest and halt the urban water utility's groundwater exploitation, because it reduced the availability of water resources for irrigation. Later, this irrigators' movement joined forces with urban citizens to oppose water utility privatization during the well-known Cochabamba Water War, engaging in both urban street protests as well as roadblocks on the city's access roads. In a similar manner, affected populations in Ecuador tried to halt dam construction through roadblocks and judicial lawsuits, the mobilization of social networks, and alliances with critical researchers and NGOs (Hidalgo-Bastidas & Boelens, 2019). Strategies and alliances are thus always dependent on the situation, relations and resources at hand, and are dynamically adjusted according to the contexts and issues at stake.

Next to open mobilizations, disputes evolve around problem and solution framings. In the Elgin Valley in South Africa, urban grass-roots organizations and marginalized populations claim water and sanitation service from responsible authorities, not only demanding better living conditions but also challenging how farmers – in line with the above-mentioned cultural politics – depict them as water polluters jeopardizing agro-export production and the overall economy (Wessels et al., 2019). Divisions and finger-pointing are so refractory in this case that, interestingly, no clear action has been undertaken to join forces between the commercial farmers and the urban poor to reduce sanitation challenges and connected water pollution from low-income areas.

There are also more subtle ways of contesting water access and control, as well as situations in which acceptance and adjustments prevail. In Bangalore, for example, communities have 'illegally' accessed water from an urban water supply pipeline that passes through their territory ('illegal water users' being another powerful legal-cultural -political identity label). They thereby materially challenged and 'reconstructed' the urban-serving water infrastructure and their assigned role as 'non-consumers' inherent in that infrastructure (Goldman & Narayan, 2019). In a different manner, in the Guadalhorce Valley in Spain, affected communities memorialize past injustices to claim recognition for them, and participation in decision making regarding new policies that transform their hydrosocial territory (Duarte-Abadía & Boelens, 2019). Here, contestations are not merely about water access and control but also about recognition of needs, identities, knowledges and pasts that have become forgotten, intentionally sidelined or sometimes literally submerged under water – communities that are actively 'unimagined' (Nixon, 2009). Such recognition efforts and identification politics exist side by side with people trying to recover, reframe and react, in order to be able to move on. As this issue's contributions show, reactions of affected water users, both rural and urban, are multiple and ambivalent, ranging from acceptance (Damonte & Boelens, 2019), to consideration of compensation for damages and losses (Torio et al., 2019), to aiming to benefit from urban-based water projects (Bleeker & Vos, 2019), to active uprising and protest (Hidalgo-Bastidas & Boelens, 2019; Wessels et al. 2019). Frequently, multiple and seemingly contradictory reactions and coping strategies coexist inside a group or even within one person.

There is thus an impressive complexity and diversity of mobilizations, day-to-day negotiations, strategies and alliances, which continuously question, renegotiate and reconfigure rural–urban water connections and relations. It is almost never a clear-cut story of deprived rural communities versus powerful urban elites but complex and cross-cutting networks of interests, relations, resources and topics that transcend rural / peri-urban / urban categories, spaces and roles. Likewise, as Wessels et al. (2019) show, actors might be present as different water users in different places at the same time: as urban water user suffering from lack of sanitation services and consequently involuntarily co-contributing to water pollution, and as agricultural wage workers depending on high-quality irrigation water for employment. Scholarship thus needs to debunk these schematic water-user-identity labels and unpack the multidirectional, multiscalar complexities and micropolitical ecologies to understand justice and equity concerns (e.g., Horowitz, 2011; Rasch & Köhne, 2016; Rocheleau & Roth, 2007; Scott, 1990) and to point towards possibilities for positive rural–urban engagements.

Concluding remarks

Given the accelerated and changing urbanization dynamics, and the shifting transformations affecting nearby peri-urban and rural territories, it is important to move beyond a focus on the urban. Instead, our analyses and policies need to consider the multidimensional linkages and concerns that connect rural, urban, and peri-urban areas, taking into account issues related to justice, social mobilizations, and shifting political and ecological dynamics. Concurrently, there is a need for new approaches and methods to study these in more careful and systematic ways. Here we do not offer a unifying framework to do so, but rather provide a series of starting points to enable movement in that direction.

In so doing, the articles of this special issue bring forward the diversity and multifaceted nature of this topic, cutting across personal, community, regional and global scales, and involving a wide range of actors. The relationship between cities, peri-urban spaces, and surrounding rural areas cannot be understood without analyzing shifting water resources and flows in conjunction with concerns related to identities, knowledges, imaginaries and discourses. As has become clear in this introduction, these facets should be considered as both source and result of the permanently negotiated and contested reconfiguration of rural–urban hydrosocial territories. The analytical focus on water flows allows analyzing, understanding and making visible their multidimensional, multidirectional and multiscalar entanglements.

Several things have stood out from the provided analyses, which we believe point to new trends and issues that should be explored in the future. First, the different case studies have exemplified the multifaceted triggers and driving forces behind urbanization processes and resulting water supply endeavours: ranging from systematic promotion of IT development and the associated upholding of a modern green illusion detached from ecological realities in Bangalore; to agro-export production in the Elgin and Ica Valleys that demands both clean, international-standard-fulfilling irrigation water as well as low-wage labourers; to nation-building efforts and commercial tourism development in Spain. This makes clear that urban development does not proceed along a natural path of development that is replicated the world over. This type of supposition is often implicit in development and urbanization reports that highlight urbanization shifts as proceeding hand in hand with societal movement from agricultural to industrial economies, from underdevelopment to development. Considering the outlined challenges associated with accelerated urbanization in general, it should first be acknowledged and then carefully analyzed how urbanization is an actively constructed phenomenon rather than inevitable or a necessary sign of 'development'.

Second, it is imperative to explore what kind of alternative policies – not just focused on urban development – are possible, and what effects these alternative approaches might have on shifting hydro-geographies, democratic rural–urban relations and attendant equity concerns of the type we have emphasized throughout this issue. Connected to this, one future topic of research would be the particularities of linkages between rural areas and middle-sized or small cities, or providing further accent on the peri-urban and the growing 'spaces of in-between' that exist across the borders of urban and rural, informal and formal. The literature in the field has an inclination to focus on large cities; studying other 'degrees' and spaces of urbanization is equally important.

Third, besides the triggers and nature of changing rural–urban relations, the different case studies have explored a range of governance techniques that seek to steer and legitimize changing social and material entities, relations and interlinkages to shape urban–rural hydrosocial territories. Again, the diversity is striking: socio-territorial strategies of urban, peri-urban and rural actors range from consciously forceful to discursive discussions about problem framing and truth claims. Yet, one can observe that there is a decisive move, especially by the economically and politically powerful public and private actors, towards governance techniques that employ discourses about modernity, inclusion, multiculturalism and recognition policies, the Human Right to Water, the Sustainable Development Goals, et cetera. Such discourses often underpin or go hand in hand with the construction of water infrastructure, which in turn reconfigures ecological as well as socio-political relations. Again, other water governance approaches are first and foremost directed towards creating incentives to direct individuals' behaviour in desired directions, and claim specific scientific-technical truths to realize envisaged projects for securing water supply. The agro-export companies in Ica controlling information and declaring exclusive water-use-efficiency practices are one example; the technocratic approach to Bangalore's water problems, and the incentives created for desired conservation behaviour in communities in Lima's watersheds (responding to neoliberal PES governmentality) are other illustrations. The landscape of environmental governance is becoming increasingly complex and subtle, and different 'arts of governance' strategically combine (Foucault, 1991; cf. Dean, 1999; Fletcher, 2017; Li, 2007; Wynne-Jones, 2012), just as critiques, responses, mobilizations and 'counter-conducts' strategically diversify (e.g., Cadman, 2010; Hall et al., 2015; Horowitz, 2011; Rasch & Köhne, 2016; Scott, 1990; Valladares & Boelens, 2017).

A fourth point extends from the interest in water governance to include explicit focus on the role of different actors in shaping environmental governance. In particular, the role of state institutions *vis-à-vis* private actors has been raised by contributions throughout the issue. In some cases (for example in Spain and Ecuador – Duarte-Abadía & Boelens, 2019; Hidalgo-Bastidias & Boelens, 2019; both this volume), a strong state has had a central role in levelling the ground and implementing large-scale hydraulic projects such as dams for urban water supply and flood protection. Yet, the other cases tell of situations where it is not that clear what role state institutions have in the reconfiguration of hydrosocial territories, particularly *vis-à-vis* other actors: it appears to be increasingly difficult to enforce state regulations in the face of actions that are characterized by self-regulation by other (private) actors (Damonte & Boelens, 2019; Goldman & Narayan, 2019). In yet other cases, such as PES implementation in Lima (Bleeker & Vos, 2019), national policies and decisions set the base and impulse for new rural–urban engagement mechanisms, but remain importantly influenced by international organizations and water management paradigms (e.g., Boelens, Perreault, & Vos, 2018; Harris et al., 2013; Zwarteveen, 2015). In Lima but also in other cases, the municipality plays no particular role, which is interesting given that on a global scale there have recently been recurring calls for attention to the important role municipalities can take independent of national and broader state politics (a very recent example being the hundreds of mayors in the United States committing to uphold the Paris climate goals despite President Trump's announcement of withdrawal – Climate Mayors, 2018). The role or non-role of state institutions and/or the independence and site-presence of municipalities is thus an interesting and important issue to explore in the future, both in terms of its potential to shape a sustainable and equitable future, as well as in terms of potential risks associated with institutional/power vacuums that are filled or purposefully appropriated by other non-public actors that construct rural–urban hydrosocial territories according to their particular visions and interest.

Therefore, finally, even though the topic of rural–urban relations has already been on the research and policy agenda for some time, we believe that the continuously changing dynamics call for sustained, critical research efforts that take into account the new actors and governance modes and latest water management paradigms. Only with a profound understanding of on-the-ground realities and power mechanisms at work will it be possible to establish steps towards democratic, equitable and sustainable rural / peri-urban / urban relations.

Disclosure statement

No potential conflict of interest was reported by the authors.

Funding

This work forms part of the activities of the Water Justice/Justicia Hídrica Alliance (www.justiciahidrica.org) and was supported by a grant of the Social Sciences and Humanities Research Council for the International WaTERS research and training network [Grant no: 890-2013-0020].

References

Aguilera-Klink, F., Pérez-Moriana, E., & Sánchez-García, J. (2000). The social construction of scarcity. The case of water in Tenerife (Canary Islands). *Ecological Economics*, *34*(2), 233–245.

Bakker, K. (2010). *Privatizing water: Governance failure and the world's urban water crisis*. Ithaca, NY: Cornell University Press.

Bleeker, S., & Vos, J. (2019). Payment for ecosystem services in lima′s watersheds: Power and imaginaries in an urban-rural hydrosocial territory. *Water International*, *44*(2), 224–242. doi:10.1080/02508060.2019.1558809

Boelens, R., Hoogesteger, J., Swyngedouw, E., Vos, J., & Wester, P. (2016). Hydrosocial territories: A political ecology perspective. *Water International*, *41*(1), 1–14. doi:10.1080/02508060.2016.1134898

Boelens, R., Perreault, T., & Vos, J. (Eds.). (2018). *Water justice*. Cambridge, UK: Cambridge University Press.

Cadman, L. (2010). How (not) to be governed: Foucault, critique, and the political. *Environment and Planning*, *28*(3), 539–556. doi:10.1068/d4509

Climate Mayors. (2018, July 27). Retrieved from http://climatemayors.org

Damonte, G., & Boelens, R. (2019). Hydrosocial territories, agro-export and water scarcity: Capitalist territorial transformations and water governance in Peru's coastal valleys. *Water International*, *44*(2), 206–223. doi:10.1080/02508060.2018.1556869

Dean, M. (1999). *Governmentality. Power and rule in modern society*. London: Sage.

Duarte-Abadía, B., & Boelens, R. (2019). Colonizing rural waters. The politics of hydro-territorial transformation in the Guadalhorce Valley, Málaga, Spain. *Water International*, *44*(2), 148–168. doi:10.1080/02508060.2019.1578080

Fletcher, R. (2017). Environmentality unbound: Multiple governmentalities in environmental politics. *Geoforum*, *85*, 311–315. doi:10.1016/j.geoforum.2017.06.009

Foucault, M. (1991). Governmentality. In G. Burchell, C. Gordon, & P. Miller (Eds.), *The foucault effect: Studies in governmentality* (pp. 87–104). Chicago: University of Chicago Press.

Goldman, M., & Narayan, D. (2019). Water crisis through the analytic of urban transformation: an analysis of Bangalore's hydrosocial regimes. *Water International*, *44*(2), 95–114. doi:10.1080/02508060.2019.1578078

Hall, R., Edelman, M., Borras, S. M., Jr., Scoones, I., White, B., & Wolford, W. (2015). Resistance, acquiescence or incorporation? An introduction to land grabbing and political reactions 'from below'. *The Journal of Peasant Studies*, *42*(3–4), 467–488. doi:10.1080/03066150.2015.1036746

Harris, L. (2015). Scalar politics, networks and power in water governance. In E. Norman, C. Cook, & A. Cohen (Eds.), *Negotiating water governance: Why the politics of scale matter* (pp. 226–250). London: Ashgate.

Harris, L., Rodina, L., & Morinville, C. (2015). Revisiting the human right to water from and environmental justice lens. *Politics, Groups and Identities*, *3*(4), 660–665. doi:10.1080/21565503.2015.1080619

Harris, L. M., Goldin, J. A., & Sneddon, C. (Eds.). (2013). *Contemporary water governance in the global South: Scarcity, marketization and participation*. London, UK: Routledge.

Hidalgo-Bastidas, J. P., & Boelens, R. (2019). The political construction and fixing of water overabundance. Rural-urban flood risk politics in coastal Ecuador. *Water International*, *44*(2), 169–187. doi:10.1080/02508060.2019.1573560

Hommes, L., Boelens, R., Duarte-Abadía, B., Hidalgo, J. P., & Hoogesteger, J. (2018). Reconfiguration of hydrosocial territories and struggles for water justice. In R. Boelens, T. Perreault, & J. Vos (Eds.), *Water justice* (pp. 151–168). Cambridge, UK: Cambridge University Press.

Hommes, L., Boelens, R., Harris, L., & Veldwisch, G. J. (2019). Rural-urban water struggles: Urbanizing hydrosocial territories and the evolving connections, discourses and identities. *Water International*, *44*(2), 81–94. doi:10.1080/02508060.2019.1583311

Hoogendam, P. (2019). Hydrosocial territories in the context of diverse and changing ruralities: The case of Cochabamba's drinking water provision over time. *Water International*, *44*(2), 129–147. doi:10.1080/02508060.2019.1551711

Hoogesteger, J., Boelens, R., & Baud, M. (2016). Territorial pluralism: Water users' multi-scalar struggles against state ordering in Ecuador's highlands. *Water International*, *41*(1), 91–106. doi:10.1080/02508060.2016.1130910

Horowitz, L. S. (2011). Interpreting industry's impacts: Micropolitical ecologies of divergent community responses. *Development and Change*, *42*(6), 1379–1391. doi:10.1111/j.1467-7660.2011.01740.x

Li, T. M. (2007). *The Will to Improve. Governmentality, Development, and the Practice of Politics*. London: Duke University Press.

Lynch, B. (2013). River of contention: Scarcity discourse and water competition in highland Peru. The Georgia Journal of International and Comparative Law, *42*, 69–92.

Mehta, L. (2006). Whose scarcity? Whose property? The case of water in western India. *Land Use Policy*, *24*(4), 654–663. doi:10.1016/j.landusepol.2006.05.009

Nixon, R. (2009). Unimagined communities: Developmental refugees, megadams and monumental modernity. *New Formations*, *69*, 62–80. doi:10.3898/newf.69.03.2010

Porter, J., & Demeritt, D. (2012). Flood-risk management, mapping, and planning: The institutional politics of decision support in England. *Environment and Planning*, *44*(10), 2359–2378. doi:10.1068/a44660

Rasch, E. D., & Köhne, M. (2016). Micropolitics in resistance: The micropolitics of large-scale natural resource extraction in South East Asia. *Society & Natural Resources*, *29*(4), 479–492. doi:10.1080/08941920.2015.1086458

Rocheleau, D., & Roth, R. (2007). Rooted networks, relational webs and powers of connection: Rethinking human and political ecologies. *Geoforum*, *38*, 433–437. doi:10.1016/j.geoforum.2006.10.003

Roth, D., Zwarteveen, M., Joy, K. J., & Kulkarni, S. (2018). Water governance as a question of justice: Politics, rights and representation. In R. Boelens, T. Perreault, & J. Vos (Eds.), *Water Justice* (pp. 43–58). Cambridge, UK: Cambridge University Press.

Scott, J. C. (1990). *Domination and the arts of resistance: Hidden transcripts*. New Haven, CT: Yale University Press.

Swyngedouw, E. (2015). *Liquid power: Contested hydro-modernities in 20th century Spain*. Cambridge, MA: MIT Press.

Swyngedouw, E., & Boelens, R. (2018). "And not a single injustice remains": Hydro- territorial colonization and techno- political transformations in Spain. In R. Boelens, T. Perreault, & J. Vos (Eds.), *Water Justice* (pp. 115–133). Cambridge, UK: Cambridge University Press.

Torio, P., Harris, L., & Angeles, L. (2019). The rural-urban equity nexus of Metro Manila's water system. *Water International*, *44*(2), 115–128. doi:10.1080/02508060.2019.1560559

Valladares, C., & Boelens, R. (2017). Extractivism and rights of nature: Governmentality, 'convenient communities' and epistemic pacts in Ecuador. *Environmental Politics*, *26*(6), 1015–1034. doi:10.1080/09644016.2017.1338384

Veldwisch, G. J., Franco, J., & Mehta, L. (2018). Water grabbing: Practices of contestation and appropriation of water resources in the context of expanding global capital. In R. Boelens, T. Perreault, & J. Vos (Eds.), *Water justice* (pp. 59–70). Cambridge, UK: Cambridge University Press.

Vos, J., & Hinojosa, L. (2016). Virtual water trade and the contestation of hydrosocial territories. *Water International*, *41*(1), 37–53. doi:10.1080/02508060.2016.1107682

Warner, J. (2010). *Flood planning: The politics of water security*. London: I.B. Tauris.

Wessels, M., Veldwisch, G. J., Kujawa, K., & Delcarme, B. (2019). Upsetting the apple cart? Export fruit production, water pollution, and social unrest in the Elgin Valley, South Africa. *Water International*, *44*(2), 188–205. doi:10.1080/02508060.2019.1586092

Wynne-Jones, S. (2012). Negotiating neoliberalism: Conservationists' role in the development of payments for ecosystem services. *Geoforum*, *43*(6), 1035–1044. doi:10.1016/j.geoforum.2012.07.008

Zwarteveen, M. (2015). Regulating water, ordering society. Practices and politics of water governance. *Inaugural Lecture, University of Amsterdam*. Retrieved from https://www.un-ihe.org/sites/default/files/oratie_margreet_zwarteveen.pdf

Index

Note: **Bold** page numbers refer to tables; *italic* page numbers refer to figures and page numbers followed by "n" denote endnotes.